# Optimization of Gridshells Against Instability Considering Joints' Mechanical Performance

Mingfei Lu · Jihong Ye · Hui Li

# Optimization of Gridshells Against Instability Considering Joints' Mechanical Performance

Springer

Mingfei Lu
CCCC Tunnel Engineering Company
Limited
Beijing, China

Jihong Ye
China University of Mining and Technology
Xuzhou, China

Hui Li
CCCC Tunnel Engineering Company
Limited
Beijing, China

ISBN 978-981-96-8203-4        ISBN 978-981-96-8204-1  (eBook)
https://doi.org/10.1007/978-981-96-8204-1

This work was supported by National Science Fund for Distinguished Young Scholars, National Key Research and Development Program of China (2017YFC1500702) and Jiangsu Province Postgraduate Research and Practice Innovation Program (KYCX17_0123).

© The Editor(s) (if applicable) and The Author(s) 2025. This book is an open access publication.

**Open Access** This book is licensed under the terms of the Creative Commons Attribution-NonCommercial-NoDerivatives 4.0 International License (http://creativecommons.org/licenses/by-nc-nd/4.0/), which permits any noncommercial use, sharing, distribution and reproduction in any medium or format, as long as you give appropriate credit to the original author(s) and the source, provide a link to the Creative Commons license and indicate if you modified the licensed material. You do not have permission under this license to share adapted material derived from this book or parts of it.

The images or other third party material in this book are included in the book's Creative Commons license, unless indicated otherwise in a credit line to the material. If material is not included in the book's Creative Commons license and your intended use is not permitted by statutory regulation or exceeds the permitted use, you will need to obtain permission directly from the copyright holder.

This work is subject to copyright. All commercial rights are reserved by the author(s), whether the whole or part of the material is concerned, specifically the rights of translation, reprinting, reuse of illustrations, recitation, broadcasting, reproduction on microfilms or in any other physical way, and transmission or information storage and retrieval, electronic adaptation, computer software, or by similar or dissimilar methodology now known or hereafter developed. Regarding these commercial rights a non-exclusive license has been granted to the publisher.

The use of general descriptive names, registered names, trademarks, service marks, etc. in this publication does not imply, even in the absence of a specific statement, that such names are exempt from the relevant protective laws and regulations and therefore free for general use.

The publisher, the authors and the editors are safe to assume that the advice and information in this book are believed to be true and accurate at the date of publication. Neither the publisher nor the authors or the editors give a warranty, expressed or implied, with respect to the material contained herein or for any errors or omissions that may have been made. The publisher remains neutral with regard to jurisdictional claims in published maps and institutional affiliations.

This Springer imprint is published by the registered company Springer Nature Singapore Pte Ltd.
The registered company address is: 152 Beach Road, #21-01/04 Gateway East, Singapore 189721, Singapore

If disposing of this product, please recycle the paper.

**Competing Interests** The authors have no competing interests to declare that are relevant to the content of this manuscript.

# Contents

1 Introduction ................................................. 1
  1.1 Background of the Research ............................... 1
  1.2 The Current Research Status .............................. 3
    1.2.1 Stability Analysis of Lattice Shell Structures ........ 3
    1.2.2 The Study of Structural Vulnerability Theory ......... 5
    1.2.3 Research on the Section Size Optimization of Space Truss Structure .................................... 6
    1.2.4 Development of Spatial Structure Joint ............... 8
    1.2.5 Research on Topology Optimization of Continuum Structure .......................................... 10
  1.3 The Main Work of This Book ............................... 12
  References .................................................. 15

2 The Stability Mechanism of Single-Layer Gridshells Based on the Theory of Configuration Vulnerability ................. 21
  2.1 Joint Well-Formedness of Configuration Vulnerability Theory .... 22
  2.2 Joint Well-Formedness Under Load ......................... 24
  2.3 Relative Gradient of Joint Well-Formedness ............... 25
  2.4 Verification of Stability Tracking Strategy Using Arc Length Method ................................................... 25
  2.5 Example 1: Single-Layer Reticulated Shell with 14 m Span .... 26
    2.5.1 Full Span Uniform Load Mode ......................... 26
    2.5.2 Half Span Uniform Load Mode ......................... 27
    2.5.3 Vertex Concentrated Load Mode ....................... 30
    2.5.4 Structure Instability Mechanism and Most Unfavorable Load Mode ............................. 32
  2.6 Example 2: Single-Layer Reticulated Shell with 22 m Span .... 33
    2.6.1 Full Span Uniform Load Mode ......................... 33
    2.6.2 Half Span Uniform Load Mode ......................... 35
    2.6.3 Vertex Concentrated Load Mode ....................... 36

|     |       | 2.6.4 Structure Instability Mechanism and Most Unfavorable Load Mode | 38 |
|---|---|---|---|
|     | 2.7 | Example 3: Single-Layer Reticulated Shell with 50 m Span | 38 |
|     |       | 2.7.1 Full Span Uniform Load Mode | 39 |
|     |       | 2.7.2 Half Span Uniform Load Mode | 41 |
|     |       | 2.7.3 Vertex Concentrated Load Mode | 43 |
|     |       | 2.7.4 Structure Instability Mechanism and Most Unfavorable Load Mode | 44 |
|     | 2.8 | Chapter Summary | 45 |
|     | References | | 45 |
| 3 | **Stability Optimization of Single-Layer Lattice Shell Structure Based on Rigid Joints** | | **47** |
|     | 3.1 | Stability Optimization Model of Single-Layer Lattice Shell Structure | 48 |
|     |       | 3.1.1 Optimize Goals | 48 |
|     |       | 3.1.2 Optimize Variables | 49 |
|     |       | 3.1.3 Constraint Conditions | 49 |
|     | 3.2 | Stability Optimization Algorithm for Single-Layer Lattice Shell Structure | 51 |
|     |       | 3.2.1 Canonical GA | 51 |
|     |       | 3.2.2 Guided GA | 52 |
|     | 3.3 | Optimization Example 1: 22 m Span Single-Layer Reticulated Shell | 55 |
|     |       | 3.3.1 Basic Information of Structure | 55 |
|     |       | 3.3.2 Stable Optimization by Standard Genetic Algorithm | 55 |
|     |       | 3.3.3 Guided Genetic Algorithm for Solving Stable Optimization | 60 |
|     |       | 3.3.4 Comparison of Optimization Algorithms | 62 |
|     | 3.4 | Optimization Example 2: 50 m Span Single-Layer Reticulated Shell | 64 |
|     |       | 3.4.1 Basic Information of Structure | 64 |
|     |       | 3.4.2 Stable Optimization by Standard Genetic Algorithm | 66 |
|     |       | 3.4.3 Guided Genetic Algorithm for Solving Stable Optimization | 67 |
|     |       | 3.4.4 Comparison of Optimization Algorithms | 70 |
|     | 3.5 | Optimization Example 3: 80 m Span Single-Layer Reticulated Shell | 72 |
|     |       | 3.5.1 Basic Information of Structure | 72 |
|     |       | 3.5.2 Guided Genetic Algorithm for Solving Stable Optimization | 73 |
|     | 3.6 | Chapter Summary | 74 |
|     | References | | 76 |

## 4 Stability Optimization and Collapse Resistance Verification of Large Single-Layer Lattice Shell Structure ............ 79
- 4.1 Large-Span Single-Layer Gridshells ............ 79
  - 4.1.1 Model Structure Design ............ 79
  - 4.1.2 Model Shaking Table Collapse Test ............ 81
- 4.2 Stability Optimization of Large Single-Layer Lattice Shell Structure ............ 85
  - 4.2.1 Model 1 (Normal Model) ............ 85
  - 4.2.2 Model 2 (Weak Model) ............ 87
  - 4.2.3 Comparison of Stable Optimization of Two Models ............ 93
- 4.3 Collapse Resistance of Model 1 After Stable Optimization ............ 94
  - 4.3.1 Shaking Table Test Process of Model 1 Initial Structure ............ 94
  - 4.3.2 Collapse Resistance of Model 1 After Stable Optimization ............ 96
- 4.4 Collapse Resistance of Model 2 After Stable Optimization ............ 97
  - 4.4.1 Shaking Table Test Process of Model 2 Initial Structure ............ 97
  - 4.4.2 Collapse Resistance of Model 2 After Stable Optimization ............ 99
- 4.5 Chapter Summary ............ 99
- References ............ 101

## 5 Topology Optimization Design of Joints of Single-Layer Gridshells ............ 103
- 5.1 Basic Components of Space Structure Joints ............ 104
- 5.2 Topology Optimization of Space Structure Joints Based on Rotational Stiffness ............ 105
  - 5.2.1 Joint Construction Design ............ 105
  - 5.2.2 Topology Optimization Model for Joint Nucleus Rotational Stiffness ............ 108
  - 5.2.3 Key Technologies for Joint Nucleus Topology Optimization (Application of Equivalent Joint Loads) ............ 110
  - 5.2.4 Joint-Core Topology Optimization Algorithm ............ 112
  - 5.2.5 Comparison of Joints (Mechanical Properties of Welded Hollow Ball Joints) ............ 112
  - 5.2.6 Determining the Optimal Topology for Maximum Stiffness in a Fixed-Quality Structure ............ 113
  - 5.2.7 Optimal Topology Design for Minimum Stiffness and Mass of Fixed Joints ............ 115
- 5.3 Topology Optimization Based on Safety Performance of Joints ............ 118
  - 5.3.1 Joint Safety Performance Indicators ............ 118
  - 5.3.2 Physical Meaning of Joint Security Indicators ............ 119
  - 5.3.3 Topology Optimization Model for Joint Safety Performance ............ 122
  - 5.3.4 Sensitivity Coefficient for Units ............ 122

|  |  | 5.3.5 | Topology Optimization Algorithm for Joint Safety Performance | 123 |
|---|---|---|---|---|

        5.3.6   Optimized Joints Based on Security Performance ......... 125
  5.4   Chapter Summary .......................................... 130
References ................................................... 133

**6 Optimization of Stability of Single-Layer Gridshells Considering Joint Stiffness** ......................................... 135
  6.1   Damage Theory Considering Joint Stiffness .................... 136
        6.1.1   Modifying the Stiffness Matrix of a Rectangular Element .............................................. 136
        6.1.2   The Degree of Freedom in the Joint Configuration Considering the Joint Stiffness ....................... 138
        6.1.3   The Mechanism of Instability in Mesh Structures with Consideration of Joint Stiffness ................... 139
  6.2   The Influence of Joint Rotational Stiffness on the Stability Performance of Single-Layer Gridshells ....................... 139
        6.2.1   Dimensionless Torsional Stiffness and Stable Bearing Capacity ............................................ 139
        6.2.2   The Effect of Rotational Stiffness on Stable Bearing Capacity ............................................ 141
        6.2.3   Range of Acceptable Joint Rotational Stiffness ........... 143
  6.3   Title Stable Optimization of Single-Layer Gridshells with Consideration of Joint Stiffness ......................... 145
        6.3.1   Optimization Goals ................................... 145
        6.3.2   Optimizing Variables ................................. 145
        6.3.3   Constraints ......................................... 146
        6.3.4   Optimization Algorithm .............................. 147
  6.4   Optimization Example 1: Single-Layer Gridshell with 22 m Span ..................................................... 152
        6.4.1   Stable Optimization Considering Joint Stiffness .......... 152
        6.4.2   The Anti-collapse Performance of Structures with Stable Optimized Stiffness After Joint Consideration ....................................... 157
  6.5   Optimization Example 2: Single-Layer Gridshell with 50 m Span ..................................................... 161
        6.5.1   Stable Optimization Considering Joint Stiffness .......... 161
        6.5.2   The Anti-collapse Performance of Structures with Stable Optimized Stiffness After Joint Consideration ....................................... 165
  6.6   Optimization Example 3: Single-Layer Gridshell with 80 m Span ..................................................... 168
        6.6.1   Stable Optimization Considering Joint Stiffness .......... 168

|     | 6.6.2 | The Anti-collapse Performance of Structures with Stable Optimized Stiffness After Joint Consideration | 171 |
| --- | --- | --- | --- |
| 6.7 | Chapter Summary | | 176 |
| References | | | 177 |

# 7 Conclusion and Prospects ... 179
## 7.1 Main Conclusions ... 179
## 7.2 Future Outlook ... 184

# Chapter 1
# Introduction

**Abstract** This section begins by presenting the background of the research topic, followed by a review of the current state of research in relevant fields, including the stability of spatial grid structures, structural vulnerability theory, size optimization of spatial grid structures, the development of new joints for spatial structures, and topology optimization of continuum structures. Finally, the main contributions of this study are outlined.

## 1.1 Background of the Research

Due to its elegant shape, economic rationality, and excellent space-spanning ability, single-layer lattice shell structures are widely used in large public buildings, such as stadiums, convention centers, and airports. For example, the Laoshan Velodrome and the Nagoya Dome of the Beijing Olympics. Many space lattice shell structures have also become local landmarks and have a huge social impact. The lattice shell structure is made of lightweight high-strength materials, which provides good seismic performance; the connection of components meets certain geometric topological relationships, which enables the ability to span large spaces.

With the progress of structural technology, the span of more than 150 m lattice shell is not individual. The span of Nagoya Dome has reached 180 m. At the same time, with the increase of span, the stability of shell structure is becoming more and more prominent. In 1963, a 93.5 m span single layer lattice shell roof in Bucharest completely collapsed after a heavy snow, belongs to the stability of lattice shell. This structural design accident made engineers realize the importance of lattice shell stability [1]. In 1993, the composite lattice shell roof of the circular coal bunker of the coal washing plant in Changcun Mine in Shanxi Province was completely flipped over because of instability in the construction process. Since then, the stability of large-span single layer lattice shell structure has been highly valued at home and abroad. The research results and engineering experience at the present stage show that the stability of lattice shell structure has gone beyond the strength and stiffness problems and become the controlling factor in the design of lattice shell structure

[2], that is, the ultimate bearing capacity of lattice shell is generally determined by the stable bearing capacity. Therefore, improving the stability of single layer lattice shell structure has important significance and value for improving the structural safety reserve, realizing a more economical and reasonable structural form, and excavating the space spanning capacity of single layer lattice shell structure.

Single layer lattice shell structure is composed of two parts, respectively, the rod and the joint of the rod. The current design method of single layer lattice shell structure is a two-stage design method. The first stage is based on the structure form and load conditions, assuming that the joint is an ideal rigid joint [3], the rod design is carried out first, and then the joint design is carried out by the rod end reaction force; the second stage is based on the first stage, the arc length method is used to calculate the overall stable bearing capacity of the structure, and check whether the stable bearing capacity of the structure meets the design requirements. In the first stage, the joints in the single layer lattice shell structure are assumed to be rigid joints, which greatly simplifies the analysis and design process of the lattice shell structure. But the joints in the actual structure are obviously not ideal rigid joints; with the advancement of the industrialization of the building industry in China, some new joints emerged are semi-rigid joints between rigid and hinged. Generally assuming that the semi-rigid joints are ideal rigid joints will overestimate the stable bearing capacity of the lattice shell structure, making the design less safe. Existing researches [4, 5] have shown that, considering the joint stiffness will reduce the structural stability bearing capacity by 30–40%. In the second stage, if the structural stability check fails, the current conditions can only rely on experience to strengthen some parts of the structure, and there is no specific guidance method at present. At the same time, the first stage is the design of the component level based on the assumption of rigid joints, which does not involve the overall stability of the structure; the second stage is the checking of the overall stability of the structure, which does not involve the components. Therefore, there are two internal contradictions in the current design method: (1) the stability of the single layer lattice shell design can only be considered through the checking in the design process, which separates the internal relationship between the component design and the overall stability of the structure, and inevitably leads to circular design and checking; (2) the current design method and stability checking are based on the assumption of rigid joints, which can neither accurately consider the stiffness of the joints, nor hinder the use and promotion of the new semi-rigid joints. The existing internal contradictions and emerging new joints will inevitably give rise to new design methods of single layer lattice shell structure, so as to improve the design efficiency, fully tap the anti-instability ability of single layer lattice shell structure, and use and promote the new joints.

## 1.2 The Current Research Status

### *1.2.1 Stability Analysis of Lattice Shell Structures*

Stability problem is a frontier and ancient problem in structural mechanics, which can be traced back to the early research of Euler on column stability in 1744. For continuous shell problems, they were systematically carried out by Lorenz [6], Timoshenko [7] and others in the early twentieth century. In 1939, Carmen and Qian [8] obtained the approximate results of spherical shell stability bearing capacity for the first time through nonlinear analysis, laying the foundation for the subsequent continuous shell stability research [1]. The stability problem of single-layer lattice shell structure was first studied by Kloppel and Schardt [9] in 1962. Early on, due to the lack of computing tools, for a long time, people had to resort to the quasi-shell method, transforming the discrete lattice shell into a continuous shell structure, and solving the structural stability bearing capacity through the analytical method in the continuous shell. Using this method, Wright [10], Hu et al. [11] proposed the approximate calculation formula of lattice shell structural stability bearing capacity. But this method has great limitations: for different problems, it is necessary to assume the possible form of instability and make corresponding approximate assumptions; the actual lattice shell is not a continuous uniform shell, but a discrete rod system. Until the 1970s, with the increasing development and wide application of computers, discretization methods have gradually become the mainstream tools in structural stability analysis. Nonlinear finite element method and beam-column element method have become the most commonly used discretization analysis methods [12]. In 1973, Oran [13] derived the beam-column element stiffness matrix expressed by a stable function, where the element internal force and displacement are both 6, with fewer calculation variables. The nonlinear finite element method is relatively simple in derivation, but the number of the element internal forces and displacements are both 12, with more calculation variables. However, the finite element model can conveniently consider such factors as initial defects and residual stresses. No matter what kind of discretization calculation model, when the structure is loaded to the stable critical point, the structural stiffness matrix approaches singularity and iterative divergence, and the traditional calculation method cannot track the behavior of the structure after the critical point. In 1979, the arc length method proposed by Riks [14] can successfully pass the critical point and track the post-buckling behavior of the structure. After the continuous development of Crisfield [15, 15], Papadrakakis [17] and Meek [18], the spherical arc length method and cylindrical arc length method were proposed successively. The arc length method is considered by Ragon and Gurdal [19] as one of the most effective calculation methods to cross the critical point and track the post-buckling path. In the 1990s, Academician Shen Shizhao further proposed the refined beam-column element considering bending coupling, large deformation and large turning angle, and established the refined model of lattice shell structure stability analysis [1]. Based on the element, the flexible balance path tracking strategy was used to conduct large-scale parametric analysis of many large and complex lattice shells

[20], and some of the conclusions obtained have been compiled into the JGJ7-2010 Technical Specification for Space Grid Structures [3].

For more than 30 years, scholars have made in-depth research on the stability of lattice shell structures, and have achieved fruitful results in local buckling, rod instability, elasto-plastic stability, joint semi-rigidity and so on. Gioncu [21] systematically reviewed the history of lattice shell stability research, summarized various instability modes of lattice shell structures, and studied the development and propagation of local instability. Liew et al. [22] proposed a rod mechanical model from the perspective of rod instability, revealing the relationship between rod buckling and overall structural stability. Wei et al. [23] introduced rod defects to induce rod instability, and proposed a stability analysis method of lattice shell structure considering rod instability. Professor Fan Feng's research group [24, 25] defined the rod instability discrimination criteria, and studied the effect of rod instability on the overall stability of the structure. Yan et al. [26] studied the coupling relationship between rod instability and overall instability, and clarified the coupling modes of two kinds of instability. To consider material nonlinearity, Nee and Haldar [27], Suzuki et al. [28] and Papadrakakis [29] analyzed the elasto-plastic stability of lattice shell structures; Cao et al. [30, 31] and Feng et al. [32] further considered the coupling relationship between geometric nonlinearity and material nonlinearity, and compared and analyzed the difference between elastic instability and elasto-plastic instability.

Since the 1980s, foreign scholars have tended to refine the analysis of the stability of single-layer lattice shell structures and began to consider the impact of joint semi-rigidity on the stability performance of the structure. The joints in the actual single-layer lattice shell structure are semi-rigid joints between completely hinged and completely rigid. In 1987, Fathelbab [33] first investigated the impact of Mero joint semi-rigidity on the stability performance of lattice shell structures from the aspects of experimental and theoretical research. Murakami [34] investigated the impact of joint rigidity on the elastic buckling performance of lattice shells under dead weight load. Kato et al. [35, 36] and Chan [37] respectively proposed two bar mechanical models considering joint rigidity, and used the bar mechanical model with elastic joints to investigate the weakening of joint rigidity on the stability bearing capacity of the structure. Mohammad [38] studied the stiffness characteristics of various joints in detail through experimental and numerical simulation methods, and studied the impact of joint semi-rigidity on the stability of lattice shell structures with different spans and different heights. Lopez et al. [39, 40] conducted experimental and numerical simulations on the stiffness characteristics of bolt ball joints and simple lattice shell structures equipped with such joints, and proposed a rapid evaluation method for the stability bearing capacity of lattice shell structures considering joint semi-rigidity [41]. With the rapid development of China's spatial structure, the impact of joint stiffness on the stability performance of large actual lattice shells has received widespread attention from academia and engineering circles. Luo et al. [42] pointed out in 1995 that the influence of joint stiffness on the stability of lattice shell structure cannot be ignored. Guo and Shen [5] studied the overall stability of the single-layer spherical lattice shell of Shanghai International Conference Center, considering the joint stiffness, and pointed out that the overall stability bearing capacity of the structure will

## 1.2 The Current Research Status

decrease by about 30% after considering the joint semi-rigidity. Lei et al. [43] simulated the joint with shell elements in the lattice shell finite element model, conducted refined simulation, and studied the influence of joint stiffness on the stability of lattice shells qualitatively with reference to the current design specifications. Li [44] introduced the joint stiffness into the continuous analysis method, which is entiled as quasi-shell method, and proposed a revised quasi-shell method considering the joint stiffness. Using commercial finite element software, scholars quantitatively studied the influence of joint stiffness on K8 lattice shell structure [45, 46], Schwedler lattice shell [47], cylindrical lattice shell [48], cable-supported lattice shell [49]. In the past decade, with the increasingly deep understanding of joint stiffness and the refinement of structural analysis and design, the stiffness characteristics of many existing joints have been widely studied. In 2009, Professor Feng et al. [50] studied the stiffness characteristics of semi-rigid bolted ball joints and their influence on the elastoplastic stability of single-layer K8 lattice shells. Ma et al. [51] studied the stiffness characteristics of Socket joints and their influence on the elastoplastic stability of lattice shells. Professor Han et al. [4] studied the stiffness of welded hollow spherical joints and their effects on the stability of single-layer lattice shells, and pointed out that semi-rigid joints could reduce the structural stability by 30–35%. Dong [52] studied the effects of hub joint stiffness on the stability of lattice shells. However, current researches mainly focus on the measurement of joint stiffness and the analysis of the effects of joint stiffness on the stability of lattice shells.

### 1.2.2 The Study of Structural Vulnerability Theory

In 1991, Professor Blockley's research group at Bristol University in the UK proposed the theory of configuration vulnerability [53–56]. The theory of configuration vulnerability regards a building structure as a system, with members as the system units. And this theory uses joint well-formedness as a physical quantity to measure the influence of members (units) on the structure (system). The theory defines vulnerability as the ratio of the structural damage consequences to the damage causing such consequences. The theory takes joint well-formedness as an indicator and comprehensively applies the hierarchical clustering method in graph theory and system theory to study the vulnerability of the structure. The specific process is as follows: the joint well-formedness is used to represent the connection degree of the structure, and the connection and combination mode between components is studied based on the structure topology configuration, and the hierarchical relationship model of the structure is established through the clustering process. Then the weak area in the structure is determined by the joint well-formedness, and the damage event is imposed on the weak area, and the hierarchical model is declustered to identify various failure modes of the structure, and the internal weak area of the structure is identified according to the vulnerability evaluation index. The theory of configuration vulnerability focuses on the configuration characteristics of the structure itself, and the weak area identified is independent of the external load, and the analysis results are universally applicable, so the theory has been widely used in the continuous collapse analysis of the pole-system structure [57]. Agarwal et al. [58] first applied this method to the three-dimensional frame structure. Pinto et al. [59] and Blockley et al. [60] combined this method with risk assessment and proposed a risk assessment method of vulnerability failure mode. Murta et al. [61] conducted vulnerability analysis on two traditional Portuguese wooden roofs, calculated the vulnerability index of various failure modes, and verified the applicability of structural vulnerability analysis method by comparing with real failure modes. Galvan and Agarwal [62] applied the configuration vulnerability theory to the robustness analysis of large-scale infrastructure. Tsinidis et al. [63] applied the configuration vulnerability theory to the risk assessment of large-scale natural gas pipelines. Professor Ye [64–66] expanded the scope of research to long-span space grid structure, combined with shaking table collapse experiment, revealed the collapse mechanism of single-layer spherical lattice shell with configuration vulnerability theory, and carried out the optimization of single-layer lattice shell structure collapse. Professor Ye et al. [67] used the tension and compression rod model to analyze the light steel keel shear wall structure, analyzed and predicted its collapse mode with configuration vulnerability theory, and verified it with shaking table experiment, further expanding the research scope of configuration vulnerability theory.

However, the classical configuration vulnerability theory only targets at the structure itself and cannot consider the load. In order to consider the load factor, Professor Zhu and Professor Ye [64] redefined the joint well-formedness, taking the reciprocal of the joint displacement under the load as the joint well-formedness. England et al. [68] conducted Pushover analysis on the failed structure under vertical load, defined disaster potential, and attempted to establish the connection between the configuration vulnerability theory and vertical load. On the other hand, the classical configuration vulnerability theory can only study the structure with ideally articulated or ideally rigid joints, and cannot consider the influence of semi-rigid joints.

### 1.2.3 Research on the Section Size Optimization of Space Truss Structure

In order to improve the service performance and economic indicators of lattice shell structure, scholars from all over the world have done a lot of research on the size optimization of lattice shell structure [69, 70]. In literature [71–74], the linear buckling load is used to represent the structural stability, and the stability problem is used as the optimization constraint condition. The linear buckling load can not consider the geometric nonlinearity of lattice shell structure, and often overestimates the actual stable bearing capacity. Pyrz [75] expresses the structural stable bearing capacity from the perspective of energy, and uses the enumeration method to optimize the simple articulated rod lattice shell. The large span single-layer lattice shell structure generally adopts beam-column elements, and the number of bars is up to thousands.

## 1.2 The Current Research Status

The enumeration method can not be applied to the actual structure. Sedaghati and Tabarok [76] express the overall structural stability with potential energy, and take the stability problem as the constraint condition. With other structural responses as the optimization goal, the size optimization of the simple lattice shell with fewer degrees of freedom is carried out. But for the actual lattice shell structure with a large number of degrees of freedom, the applicability of this method is still worth discussing. In the optimization model, the stability problem is used as the constraint condition, which can only ensure that the stable bearing capacity of the optimized structure is not lower than the given limit, and can not realize the optimization of the stable bearing capacity of lattice shell structure. Therefore, Kamat et al. [77] directly took the structural stability bearing capacity as the optimization target, and optimized the simple hinged rod structure steadily by the potential energy standing value principle under the premise of given steel consumption. Talaslioglu [78, 79] constructed the fitness function of the stable critical load obtained by the arc length method, and optimized the stable bearing capacity and steel consumption of the lattice shell structure with multi-objective optimization by genetic algorithm. However, the stability of the structure is represented by the stable critical load, which requires nonlinear arc length tracking and consumes a lot of calculation time. The current optimization model directly taking stability as the target is limited to simple hinged structures, and the calculation efficiency is not ideal, which is difficult to be applied to large-scale practical engineering.

In order to solve the structural optimization model, the corresponding optimization algorithm has been developed. According to the theoretical basis of the algorithm, it can be divided into three categories [80–82], namely optimization criterion method, mathematical programming method and heuristic optimization algorithm. The criterion method is from the perspective of practical engineering, proposes the criteria that the structure should meet when it reaches the optimal, and then finds the solution that meets these criteria by iterative method. However, the criterion method needs to construct different optimal iterative criteria, and its generality is limited. The mathematical programming method is to reduce the optimization problem to the mathematical programming problem, and solve it by using the mathematical programming method. Its disadvantage is that for the high-dimensional optimization problem, the calculation is large and the convergence speed is slow. Since the 1980s, heuristic optimization algorithm has made great progress. Heuristic optimization algorithm is an advanced algorithm that can search efficiently in the global range. Common heuristic algorithms include genetic algorithm [83], particle swarm optimization [84], ant colony algorithm [85], tabu algorithm [86], annealing algorithm [87], etc. In the field of architectural structure optimization, genetic algorithm is one of the most widely used heuristic optimization algorithms. Genetic algorithm is an optimization algorithm that simulates the genetic and evolution of species in nature, and was first proposed by Professor Holland in the United States in the 1960s [83]. A feasible solution of the optimization model is called an individual, and the set of current individuals is called a population. The binary code corresponding to an individual is called a chromosome, and the binary code in a chromosome is called a gene. The population in nature will reproduce and produce offspring. In the reproductive process, some genes mutate, and the offspring will have new forms. The selection operation is used to achieve the survival of the fittest, and a better generation of population is produced [83]. Genetic algorithm is simple to code and operate, and is not restricted by the continuity or derivation of the optimization model, so it is widely used in management science, machine learning, engineering design and other fields. By the 1980s, Goldberg [88] comprehensively and completely discussed the basic principles and application of genetic algorithm, and established the basic framework of genetic algorithm. Davis [89] then edited the genetic algorithm manual, promoting the application and promotion of genetic algorithm. In 1986, Goldberg and Samtani [90] first used genetic algorithm to optimize the plane truss. In 1992, Rajeev and Krishnamoorthy [91] and Hajela and Lin [92] developed genetic algorithm for discrete variables almost simultaneously to optimize the section of the rod structure. With the wide application of genetic algorithm in the field of structural optimization, the number of optimization variables increases, and the optimization problem becomes increasingly complex. Lin and Hajela [93] improved the crossing mechanism in genetic algorithm, improved the search efficiency of genetic algorithm in solving multivariate problems, and compiled the EVOLVE program. Azad et al. [94] took the minimum steel consumption as the optimization goal, applied the principle of virtual work to identify key components, and then guided the random search process, and proposed a design-oriented heuristic optimization method. In order to combine structural optimization and structural design, Kicinger et al. [95] summarized the achievements of evolutionary algorithm in the field of structural design and structural innovation, and pointed out that evolutionary algorithm is an important way to achieve structural innovation. However, for high-dimensional optimization problems, genetic algorithm has problems such as large population, slow convergence speed in the late stage, and the selection of optimization parameters depends on experience.

### *1.2.4 Development of Spatial Structure Joint*

Joint is an important part of the single-layer lattice shell structure. As the intersection point of the bars, the performance of the joint has an important impact on the performance of the entire structure, which is reflected in the following three aspects: (1) the stiffness of the joint directly affects the stiffness [37, 38], dynamic response [66, 96] and stability [33] of the entire structure; (2) the safety performance of the joint directly affects the safety performance of the entire structure. Once the joint fails, the adjacent bars will lose a certain degree of bearing capacity, which may cause the change of the force transmission path, the local damage of the structural system, or even cause the continuous failure of the entire structure [97]; (3) the connection mode of the joint (such as bolt connection, welding, casting, etc.) determines the construction technology of the entire structure. The master of spatial structure Makowski pointed out that the joint is an extremely important component of the spatial structure, and the ultimate commercial success of the entire project depends on the effectiveness and convenience of the joint [98]. Therefore, the development of efficient joints has been a hot topic in the study of spatial structure.

According to incomplete statistics, the number of existing spatial structure joints is more than 250 [98, 99], but not many of them can stand the test of time. Typical joint forms include German Mero joint, Oktaplatte joint, Spanish Orona joint, Canadian Triodetic joint, British Space Deck system, Nodus joint, American Octa Hub joint, Unistrut joint, Temcor joint, French Newbat joint, Segmo joint, Tridi 2000 joint, Italian Premit joint, etc. Chong et al. [97] and Gasii [100] systematically reviewed various types of spatial structure joints combined with a large number of engineering practices, and put forward guiding opinions on the design and development of joints. Drawing on the structure of classical joints, while considering the construction process, analogous to the component method of beam-column joints in European regulations [101], some new types of joints have been developed in recent years. Li et al. [102] from Beijing University of Technology improved the connection mode of welded hollow ball joint and rod, and developed sleeve connected welded hollow ball joint. Professor Fan Feng and his team from Harbin Institute of Technology developed semi-rigid C-type joint [103], BC joint [104], gear-bolt joint [105, 106], and carefully studied the mechanical properties of relevant joints. Professor Han et al. [107] from Tianjin University developed a hub joint with a high degree of assembly. Professor Feng et al. [108, 109] improved the sleeve joint and double-layer ring joint respectively. In order to further consider the economic performance and repairability of joints, Oh et al. [110] developed FREE joints. For the aluminum alloy joints that have emerged in recent years, many scholars have conducted sufficient research [111–114]. Weng et al. [115] systematically summarized various forms of spatial grid structure joints and pointed out the development requirements of spatial structure joints combined with engineering practice.

Another joint development method is the topology optimization method. In the joint feasible region, the continuous topology optimization is carried out to obtain the appropriate joint form. This method does not rely on existing experience and can be optimized for a given optimization target. The continuous topology optimization has been widely used in machinery, aviation, automobile and other fields, but there are few related studies in the field of civil engineering. Khalaf and Saka [116] took the lead in using the topology optimization method to optimize the design of steel joint plate in 2007. Ren and Galjaard [117] optimized the topology of the joints of an integral tension structure in The Hague, the Netherlands. Finally, under the premise that the maximum equivalent stress does not exceed the allowable stress, the optimized joint height is only half of the unoptimized joint height, and the volume reduction is up to 75%. Using the topology optimization technology and 3D printing technology, RMIT, ARUP and LEAD jointly carried out the Smart Joints plan [118], aiming to develop efficient joints, simplify the processing process and reduce labor costs. In 2018, Seifi et al. [119] took the maximum joint stiffness as the optimization target, adopted the BESO method, developed a new type of joint, and compared it with the joints in the actual project. The analysis showed that the joints optimized by BESO had good mechanical performance and economic indicators. Zhao et al.

[120, 121] carried out the topology optimization design of cable-stayed structure joints combined with advanced manufacturing methods. Feng et al. [122] developed an assembled joint of hexagon space grid structure in the outer ring.

Unlike the joints of double-layer grid structure, cable structure and integral tension structure, the joints of single-layer lattice shell structure must not only transmit axial force, but also transmit bending moment at the same time. Unlike the joints of frame structure, the joints of single-layer lattice shell structure have no obvious principal axis and weak axis. Therefore, as a rigid joint, the joints of single-layer lattice shell structure are subjected to more complex forces and have more demanding design requirements. With the development of diversified shapes of space grid structure, the geometric adaptability and personalized requirements of joints are higher, and the traditional design and development methods are difficult to fully meet the design requirements of the new era.

### 1.2.5 Research on Topology Optimization of Continuum Structure

The optimization of engineering structure is divided into three levels, namely size optimization, shape optimization and topology optimization, as shown in Fig. 1.1. Size optimization is carried out with component size as the optimization variable on the basis of keeping shape and topology connection unchanged; shape optimization is carried out with boundary position as the optimization variable on the basis of keeping topology connection unchanged; topology optimization is to use scientific methods to seek how limited materials are distributed in space so as to obtain the best structural performance [123]. Topology optimization is the highest level and the most difficult.

In the design domain $\Omega$, the optimization variable of topology optimization is the state $\rho$ of unit $x$, where 0 generally indicates that the unit is empty and 1 indicates that the unit exists. The function F is used to represent the target response of the structure, and $V_0$ is the upper limit of the given volume. Then, the general topology optimization model is:

$$\min: F = \int_\Omega f(\mathbf{u}(\rho), \rho)dV$$

$$s.t.: \int_\Omega \rho(x)dV - V_0 \leq 0$$

$$: C_i(\mathbf{u}(\rho), \rho) \leq 0, i = 1, 2 \ldots M$$

$$: \rho(x) = 0 \ or \ 1 \tag{1.11}$$

In the equation, **u** is the state field, which is generally the displacement field in structural optimization; $M$ is the number of constraint conditions.

In order to solve the optimization model in, the commonly used continuum topology optimization methods mainly include homogenization method, variable density method, boundary change method, progressive structure optimization method, etc.

(1) Homogeneity method

The homogenization method was proposed by Bendsøe and Kikuchi [124] in 1988. The homogenization method believes that the structure is composed of micro-units with cavities, and the structural topological relationship is represented by the size and spatial orientation of cavities as optimization variables. By establishing the relationship between the properties of micro-units and cavities, the elastic modulus and density of materials are represented by cavities. By optimizing the size and orientation of micro cavities, the topology optimization of macro-structures is realized. Although the homogenization method has defects such as large number of variables, complex iterative solution, and unsatisfactory optimization results, it is the pioneer of continuum topology optimization method and the basis of the more mature variable density method.

(2) Variable density method

The variable density method discards the cavities in the homogenization method and assumes that the unit relative density can be continuous between 0 and 1, that is, the intermediate density $\rho$ exists. In the variable density method, a variable $\rho$ can be used to express the degree of weakening of the unit by the cavities in the homogenization method. In the variable density method, the most widely used is the solid isotropic material with penalization (SIMP) [125–127]. In the SIMP method, the intermediate density unit is an isotropic material, and its elastic modulus is reduced by a penalty function. The penalty function is the cubic root of the unit relative density $\rho$. Through the penalty function, the relationship between the overall structural response and the unit relative density can be established. To solve the numerical difficulties of the SIMP method in dealing with low-density units, Stolpe and Svanberg [128] constructed another form of penalty function and proposed the RAMP method. It is precisely because of the simplicity of the variable density method that the application range of the method is the widest [129].

(3) Boundary variation method

In the first two methods, the structural topological connection is described by the state of internal units. The boundary change method represents the structural topological connection through the structural boundary surface, and obtains the reasonable structural form through the change, generation and disappearance of the surface. The boundary change method is a recently developed algorithm. Compared with other methods, the boundary change method can directly obtain smooth and accurate boundaries. The main boundary change methods are the level set method [130] and the phase field method [131].

Level set method first constructs the field function of the boundary according to the optimization objectives and constraints, and solves the Hamilton–Jacobi equation of the field function on the boundary to obtain the change velocity of the boundary to achieve the corresponding optimization objectives. However, level set method cannot adaptively generate new cavities, resulting in the optimization results depending on the initial topological configuration [129].

The idea of phase field method is similar to that of level set method, but the difference is that phase field method solves the control equation in the whole design domain (level set method only solves the control equation at the boundary), so it does not need the information about the boundary.

As a new topology optimization method, boundary variation method is getting more and more attention in academic circles. Some commercial topology optimization softwares have already started to use boundary variation method in their test versions.

(4) Progressive structural optimization method

ESO (Evolutionary Structural Optimization) was proposed by Academician Xie Yimin and Steven GP [132] in 1993. Compared with the variable density method, the state of the unit in the ESO method is only 0 or 1, and there is no intermediate density. ESO directly deletes the inefficient units in the structure to obtain the optimal structure. In 1998, on the basis of deleting the inefficient units, the introduction of efficient units into the optimization algorithm was proposed, and the bidirectional ESO method was proposed [133]. After continuous development, the BESO method has been extended in nonlinear and load-following aspects [134]. The ESO/BESO method is easy to implement, and the structure concept is clear, without requiring too complex mathematical algorithms, and has been widely used in the engineering field.

## 1.3 The Main Work of This Book

In order to solve the inherent contradiction between the current design method of lattice shell structure and the characteristics of lattice shell structure, and to use and promote the new joint, this book intends to conduct a study on the stability optimization of single-layer lattice shell structure considering the mechanical characteristics of joints. The main work is as follows:

(1) Expand the classical configuration vulnerability theory and reveal the instability mechanism of lattice shell structure

In view of the lack of external factors considered in the classical configuration vulnerability theory, it is proposed to introduce external factors (such as load, support conditions, etc.) into the configuration vulnerability theory to expand the research scope of the configuration vulnerability theory. By using the expanded configuration vulnerability theory, the instability mechanism of lattice shell structure is revealed from a new

## 1.3 The Main Work of This Book

perspective. Taking three single-layer lattice shell structures with different spans as examples, the full span uniform load, half span uniform load and vertex concentrated load are analyzed for each structure respectively to verify the instability mechanism of lattice shell structure and clarify the most unfavorable load mode.

(2) Establish a single-layer lattice shell structure stability optimization model and develop an efficient optimization algorithm

Based on the instability mechanism of lattice shell structure, and under the premise of rigid joint assumption, a lattice shell structure stability optimization model is established. The optimization model aims to reduce the instability trend of lattice shell structure and improve the structural stability bearing capacity as the direct optimization goal. With discrete bar section as the optimization variable and the design limitations specified in the specification as the optimization constraint conditions, the stability optimization model of lattice shell structure is established on the premise of given steel consumption of bar parts. In view of the large number of variables in the stability optimization model and the problem that the conventional optimization algorithm cannot solve, this paper improves the standard genetic algorithm on the basis of the standard genetic algorithm. The guided genetic algorithm can quickly solve the stability optimization model of large single-layer lattice shell structure. Finally, a small span, a medium span and a long span single-layer lattice shell structure are taken as examples to verify the optimization model and optimization algorithm.

(3) Verification of the stability optimization design method of large actual single-layer lattice shell structure and its collapse resistance performance verification

In order to further verify the applicability of the stability optimization design method, two actual large single-layer lattice shell structures used for shaking table collapse experiments are taken as examples for stability optimization design. Each lattice shell has up to 3660 optimization variables, and the value range of each variable is up to 575 candidate sections. At the same time, the support conditions, topological connection and steel consumption of the two actual lattice shell structures are completely the same, only the distribution of the bar section is slightly different. The two actual single-layer lattice shell structures are stable optimized to verify the adaptability of the stable optimization design method in large single-layer lattice shell structures;the two optimized lattice shell structures are compared with the optimization results under different initial conditions to verify the robustness of the optimization algorithm. Through numerical simulation, the two optimized lattice shell structures are planned to input the same seismic wave as the lattice shell structure before optimization, to study the anti-collapse ability of the stable optimized structure, and compare with the shaking table experimental data to verify the anti-collapse ability of the lattice shell structure after optimization.

(4) Optimization design of single-layer lattice shell structure joints

It is planned to develop new joints of single-layer lattice shell structure from the two aspects of improving the joint rotational stiffness and joint safety performance. To improve the joint rotational stiffness, the joint nuclear rotational stiffness optimization model is established, and the topology optimization method is used to optimize the joint nuclear on the premise of meeting the design requirements. At the same time, combined with the structural design, a common connection interface is designed to meet the connection requirements of different positions of the bar. To improve the joint safety performance, it is planned to define an index to evaluate the joint safety performance. Based on this index, the joint safety performance optimization model is established, and the optimization algorithm program is compiled to optimize the topology of two-dimensional joints.

(5) Optimization of stability performance of single-layer lattice shell structure considering joint stiffness. In view of the problem that the configuration vulnerability theory cannot consider the joint stiffness, the influence of joint stiffness is proposed to be considered by modifying the unit stiffness matrix, so as to further expand the configuration vulnerability theory. From the two perspectives of structural stability bearing capacity and the deformation of semi-rigid joints, the influence of joint stiffness on the stability performance of single-layer lattice shell structure is quantitatively studied, and it is verified that the configuration vulnerability theory considering joint stiffness can reveal the stability mechanism of semi-rigid joint lattice shell structure, and at the same time clarify the reasonable range of joint stiffness for single-layer lattice shell structure suitable for design. Within the reasonable range of joint stiffness, the joint research and development method proposed in this paper is used for joint optimization design, and an optimization joint library is established. In the stability optimization model, the joints and bars are proposed to be optimized variables at the same time, and a stable optimization design method considering joint stiffness is established. Finally, a small span, a medium span and a long span single-layer lattice shell structure are taken as examples to verify the stable optimization design method considering joint stiffness. The collapse resistance of the optimized lattice shell based on the rigid joint hypothesis and the optimized lattice shell considering joint stiffness are analyzed to verify the collapse resistance of the stable optimization lattice shell structure considering joint stiffness.

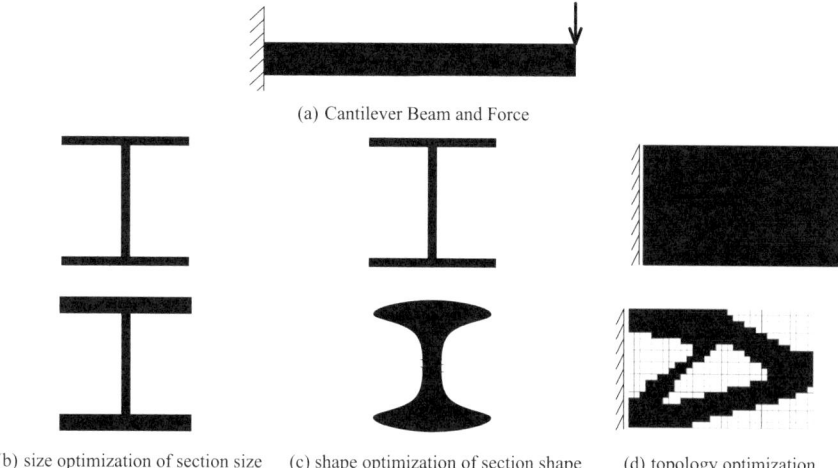

**Fig. 1.1** Various optimizations of a beam

# References

1. Shizhao S, Xin C (1999) Stability of reticulated shell structures. Science Press, Beijing
2. Shilin D, Yaozhi L, Yang Z et al (2006) Analysis, design and construction of new spatial structures. People's Communications Press, Beijing
3. Industrial standard of the People's Republic of China (2010) JGJ7-2010 technical specification for space grid structure. China Architecture and Building Press, Beijing
4. Han Q, Liu Y, Xu Y (2016) Stiffness characteristics of joints and influence on the stability of single-layer latticed domes. Thin-Walled Struct 107:514–525
5. Xiaonong G, Zuyan S (2004) Integral stability analysis of single-layer spherical lattice shell with semi-rigid joints. Sichuan Archit Sci Res 30(3):10–12
6. Lorenz R (1908) Buckling of a cylindrical shell under axial compression. Z Ver Dtsch Ing 52:1706–1713
7. Timoshenko SP (1910) Einige stabilitäts probleme der elastizitäts theorie. Zeitschrift für Mathematik und Physik 58(4):337–385
8. Karman TV, Tsien H (1939) The buckling of spherical shells by external pressure. J Aeronaut Sci 7(2):43–50
9. Kloppel K, Schardt R (1962) Zur Berechnung von Netzkuppeln. Der Stahlbau 31(5):129–136
10. Wright DT (1965) Membrane forces and buckling in reticulated shells. J Struct Div 91(1):173–202
11. Hu X (1986) Stability calculation of dome lattice shell. In: The third session of space structure academic exchange conference proceedings, vol II
12. Wu J, Zhang Q (2002) Research progress on stability of lattice shell structures. Space Struct 8(1):10–18
13. Oran C (1973) Tangent stiffness in plane frames. J Struct Div 99(6):973–985
14. Riks E (1979) An incremental approach to the solution of snapping and buckling problems. Int J Solids Struct 15(7):529–551
15. Crisfield MA (1981) A fast incremental/iterative solution procedure that handles "snap-through." Comput Struct 13(1–3):55–62
16. Crisfield MA (1983) An arc-length method including line searches and accelerations. Int J Numer Meth Eng 19(9):1269–1289

17. Papadrakakis M (1981) Post-buckling analysis of spatial structures by vector iteration methods. Comput Struct 14(5–6):393–402
18. Meek JL, Tan HS (1984) Geometrically nonlinear analysis of space frames by an incremental iterative technique. Comput Methods Appl Mech Eng 47(3):261–282
19. Ragon SA, Gurdal Z, Watson LT (2002) A comparison of three algorithms for tracing nonlinear equilibrium paths of structural systems. Int J Solids Struct 39(3):689–698
20. Shen S (1999) Stability of lattice shell structures. Chinese J Civil Eng 32(6):11–19, 25 (1999)
21. Gioncu V (1994) Buckling of reticulated shells: state-of-the-art. Int J Space Struct 10(1):1–46
22. Liew JYR, Punniyakotty NM, Shanmugam NE (1997) Advanced analysis and design of spatial structures. J Constr Steel Res 42(1):21–48
23. Tian W, Zhao Y, Dong S (2012) Stability analysis method of lattice shell structure considering bar failure. Eng Mechan 29(10):149–156
24. Feng F, Jiachuan Y, Zhenggang C (2012) Stability of lattice shell structures considering the influence of bar instability. Chin J Civil Eng 45(05):8–17
25. Fan F, Yan J, Cao Z (2012) Stability of reticulated shells considering member buckling. J Constr Steel Res 77:32–42
26. Yan J, Qin F, Cao Z et al (2016) Mechanism of coupled instability of single-layer reticulated domes. Eng Struct 114:158–170
27. Nee KM, Haldar A (1988) Elastoplastic nonlinear post-buckling analysis of partially restrained space structures. Comput Methods Appl Mech Eng 71(1):69–97
28. Suzuki T, Ogawa T, Ikarashi K (1992) Elasto-plastic buckling analysis of rigidly jointed single layer reticulated domes. Int J Space Struct 7(4):363–368
29. Papadrakakis M (1983) Inelastic post-buckling analysis of trusses. J Struct Eng 109(9):2129–2147
30. Zhenggang C, Feng F, Shizhao S (2006) Elasto-plastic stability of single-layer spherical lattice shells. Chin J Civil Eng 39(10):6–10
31. Zhenggang C, Feng F, Shizhao S (2007) Study on the elastic-plastic stability of a single-layer spherical lattice shell. Eng Mechan 24(5):17–23
32. Fan F, Yan J, Cao Z (2012) Elasto-plastic stability of single-layer reticulated domes with initial curvature of members. Thin-Walled Struct 60:239–246
33. Fathelbab FA (1987) The Effect of Joints on the Stability of Shallow Single Layer Lattice Domes. University of Cambridge
34. Murakami M (1992) Numerical analysis of elastic buckling of single-layer latticed domes under gravity load. In: Srivastava NK (ed) Innovative large span structures. IASS, Toronto, pp 576–586
35. Kato S, Mutoh I, Shomura M (1994) Effect of joint rigidity on buckling strength of single layer lattice domes. In: Spatial, lattice and tension structures. ASCE, pp 469–478
36. Kato S, Mutoh I, Shomura M (1996) Collapse of imperfect reticulated dome with semi-rigid connections. In: Chan SL, Teng JG (eds) Advances in steel structures (ICASS '96). Proceedings of international conference on advances in steel structures, Hong Kong, pp 315–320
37. Chan SL (1993) Nonlinear static and dynamic analysis of space frames with semi-rigid connections. Int J Space Struct 8(4):261–269
38. Mohammad RC (1997) Semi-rigidity of connections in space structures. University of Surrey
39. López A, Puente I, Serna MA (2007) Numerical model and experimental tests on single-layer latticed domes with semi-rigid joints. Comput Struct 85(7–8):360–374
40. López A, Puente I, Aizpurua H (2011) Experimental and analytical studies on the rotational stiffness of joints for single-layer structures. Eng Struct 33(3):731–737
41. López A, Puente I, Serna MA (2007) Direct evaluation of the buckling loads of semi-rigidly jointed single-layer latticed domes under symmetric loading. Eng Struct 29(1):101–109
42. Luo Y, Hu S, Shen Z et al (1995) Study on the influence of joint rigidity on the stability of single-layer cylindrical lattice shells with arch ribs. Spat Struct 1(2):18–22
43. Gu L, Ding M, Fu X et al (2011) Fine finite element analysis of welded single-layer spherical lattice shells. J Build Struct 32(8):42–50

44. Li L (2012) Study on the performance of lattice shell structures considering semi-rigid joints. Zhejiang University
45. Wang X, Wang F (2011) Influence analysis of joint's stiffness on spatial reticulated shell. Adv Mater Res 163–167:439–442
46. Fan F, Ma H, Cao Z et al (2011) A new classification system for the joints used in lattice shells. Thin-Walled Struct 49(12):1544–1553
47. Zhao Z, Liu H, Liang B (2017) Novel numerical method for the analysis of semi-rigid jointed lattice shell structures considering plasticity. Adv Eng Softw 114:208–214
48. Ma H, Fan F, Wen P et al (2015) Experimental and numerical studies on a single-layer cylindrical reticulated shell with semi-rigid joints. Thin-Walled Struct 86:1–9
49. Wang X, Feng R, Yan G et al (2016) Effect of joint stiffness on the stability of cable-braced grid shells. Int J Steel Struct 16(4):1123–1133
50. Fan F, Ma H, Shen S (2009) Elastic-plastic stability analysis of single-layer K8 spherical lattice shells with semi-rigid bolted ball joints. China Civil Eng J 42(2):45–52
51. Ma H, Fan F, Zhong J et al (2013) Stability analysis of single-layer elliptical paraboloid latticed shells with semi-rigid joints. Thin-Walled Struct 72:128–138
52. Dong J (2016) Study on the semi-rigidity of joints in single-layer lattice shells and its influence on the overall stability of structures. Zhejiang University
53. Wu X (1991) Vulnerability analysis of structural systems. University of Bristol
54. Wu X, Blockley DI, Woodman NJ (1993) Vulnerability of structural systems Part 1: rings and clusters. Civ Eng Syst 10(4):301–317
55. Wu X, Blockley DI, Woodman NJ (1993) Vulnerability of structural systems Part 2: failure scenarios. Civ Eng Syst 10(4):319–333
56. Lu Z, Yu Y, Woodman NJ et al (1999) A theory of structural vulnerability. Struct Eng 77(18):17–24
57. Zhao X, Yan S, Chen Y (2013) Research methods and current situation of progressive collapse in long-span spatial structures. J Build Struct 34(4):1–14
58. Agarwal J, Blockley D, Woodman N (2001) Vulnerability of 3-dimensional trusses. Struct Saf 23(3):203–220
59. Pinto JT, Blockley DI, Woodman NJ (2002) The risk of vulnerable failure. Struct Saf 24(2–4):107–122
60. Blockley DI, Agarwal J, Pinto JT et al (2002) Structural vulnerability, reliability and risk. Prog Struct Mat Eng 4(2):203–212
61. Murta A, Pinto J, Varum H (2011) Structural vulnerability of two traditional Portuguese timber structural systems. Eng Fail Anal 18(2):776–782
62. Galvan G, Agarwal J (2017) Community detection in action: identification of critical elements in infrastructure networks. J Infrastruct Syst 24(1):04017046
63. Tsinidis G, Di Sarno L, Sextos A et al (2019) A critical review on the vulnerability assessment of natural gas pipelines subjected to seismic wave propagation. Part 1: Fragility relations and implemented seismic intensity measures. Tunnell Undergr Space Technol 86:279–296
64. Zhu N, Ye J (2013) Structural vulnerability of a single-layer dome based on its form. J Eng Mech 140(1):112–127
65. Liu W, Ye J (2013) Optimization of failure modes for single-layer spherical lattice shell structures based on genetic simulated annealing algorithm. J Build Struct 34(5):33–42
66. Liu W, Ye J (2014) Collapse optimization for domes under earthquake using a genetic simulated annealing algorithm. J Constr Steel Res 97:59–68
67. Ye J, Jiang L, Wang X (2017) Seismic failure mechanism of reinforced cold-formed steel shear wall system based on structural vulnerability analysis. Appl Sci 7(2):182
68. England J, Agarwal J, Blockley D (2008) The vulnerability of structures to unforeseen events. Comput Struct 86(10):1042–1051
69. Stolpe M (2016) Truss optimization with discrete design variables: a critical review. Struct Multidiscip Optim 53(2):349–374
70. Saka MP, Geem ZW (2013) Mathematical and metaheuristic applications in design optimization of steel frame structures: an extensive review. Math Probl Eng 2013:1–33. https://doi.org/10.1155/2013/271031

71. Khot NS, Venkayya VB, Berke L (1976) Optimum structural design with stability constraints. Int J Numer Meth Eng 10(5):1097–1114
72. Khot NS (1983) Nonlinear analysis of optimized structure with constraints on system stability. AIAA J 21(8):1181–1186
73. Levy R (1994) Optimal design of trusses for overall stability. Comput Struct 53(5):1133–1138
74. Kočvara M (2002) On the modelling and solving of the truss design problem with global stability constraints. Struct Multidiscip Optim 23(3):189–203
75. Pyrz M (1990) Discrete optimization of geometrically nonlinear truss structures under stability constraints. Struct Optim 2(2):125–131
76. Sedaghati R, Tabarrok B (2000) Optimum design of truss structures undergoing large deflections subject to a system stability constraint. Int J Numer Meth Eng 48(3):421–434
77. Kamat MP, Khott NS, Venkayyat VB et al (1984) Optimization of shallow trusses against limit point instability. AIAA J 22(3):403–408
78. Talaslioglu T (2012) Multiobjective size and topolgy optimization of dome structures. Struct Eng Mech 43(6):795–821
79. Talaslioglu T (2013) Global stability-based design optimization of truss structures using multiple objectives. Sadhana 38(1):37–68
80. Bofang Z (1984) Principles and applications of structural optimization design. Water Resources and Electric Power Press, Beijing
81. Qian LX (1983) Optimization design of engineering structure. Water Resources and Electric Power Press, Beijing
82. Glover FW, Kochenberger GA (2006) Handbook of metaheuristics. Springer Science & Business Media
83. Holland JH (1992) Adaptation in natural and artificial systems: an introductory analysis with applications to biology, control, and artificial intelligence. MIT Press, Boston
84. Kennedy J, Eberhart R (1995) Particle swarm optimization (PSO). In: Proceedings of IEEE international conference on neural networks. IEEE, Perth, pp 1942–1948
85. Dorigo M, Gambardella LM (1997) Ant colony system: a cooperative learning approach to the traveling salesman problem. IEEE Trans Evol Comput 1(1):53–66
86. Glover F (1989) Tabu search—part I. ORSA J Comput 1(3):190–206
87. Kirkpatrick S, Gelatt CD, Vecchi MP (1983) Optimization by simulated annealing. Science 220(4598):671–680
88. Goldberg DE (1989) Genetic algorithms in search, optimization and machine learning. Addison-Wesley Longman Publishing Co., Boston
89. Davis L (1991) Handbook of genetic algorithms. Van Nostrand Reinhol, New York
90. Goldberg DE, Samtani MP (1986) Engineering optimization via genetic algorithm. In: Will KM (ed) Proceeding of the 9th conference on electronic computation. ASCE, New York, pp 471–482
91. Rajeev S, Krishnamoorthy CS (1992) Discrete optimization of structures using genetic algorithms. J Struct Eng 118(5):1233–1250
92. Hajela P, Lin CY (1992) Genetic search strategies in multicriterion optimal design. Struct Optim 4(2):99–107
93. Lin CY, Hajela P (1994) Design optimization with advanced genetic search strategies. Adv Eng Softw 21(3):179–189
94. Azad SK, Hasançebi O, Saka MP (2014) Guided stochastic search technique for discrete sizing optimization of steel trusses: a design-driven heuristic approach. Comput Struct 134:62–74
95. Kicinger R, Arciszewski T, De Jong K (2005) Evolutionary computation and structural design: a survey of the state-of-the-art. Comput Struct 83(23–24):1943–1978
96. Kato S, Murata M (1997) Dynamic elasto-plastic buckling simulation system for single layer reticular domes with semi-rigid connections under multiple loadings. Int J Space Struct 12(3–4):161–172
97. Fan C, Yang S, Luan HQ (2011) Research progress and practice of spatial structure joint design. J Build Struct 32(12):1–15

## References

98. Makowski ZS (2002) Development of jointing systems for modular prefabricated steel space structures. In: Proceedings of the international symposium. IASS, Warsaw, pp 17–41
99. Liu XL (2000) Review of spatial structure joints at home and abroad. In: Proceedings of the 9th academic conference on spatial structures. Bridge and Structural Engineering Branch of Chinese Society of Civil Engineering, Xiaoshan, pp 10–18
100. Gasii GM (2017) Structural and design specifics of space grid systems. Sci Tech 16(6):475–484
101. European Committee for Standardization (CEN) (2005) Eurocode 3: design of steel structures—Part 1–8: Design of joints
102. Li S, Li X, Xue S et al (2016) Strength performance of threaded-sleeve connector under axial force. Adv Struct Eng 19(7):1177–1189
103. Fan F, Ma H, Jiang X (2016) Optimization and experimental study on rotational performance of semi-rigid C-type joints in spatial structures. J Build Struct 37(3):134–140
104. Ma H, Ren S, Fan F (2016) Experimental and numerical research on a new semi-rigid joint for single-layer reticulated structures. Eng Struct 126:725–738
105. Ma H, Ma Y, Yu Z et al (2017) Experimental and numerical research on gear-bolt joint for free-form grid spatial structures. Eng Struct 148:522–540
106. Ma Y, Ma H, Yu Z et al (2018) Experimental and numerical study on the cyclic performance of the gear-bolt semi-rigid joint under uniaxial bending for free-form lattice shells. J Constr Steel Res 149:257–268
107. Han Q, Liu Y, Zhang J et al (2017) Mechanical behaviors of the Assembled Hub (AH) joints subjected to bending moment. J Constr Steel Res 138:806–822
108. Feng R, Liu F, Yan G et al (2017) Mechanical behavior of Ring-sleeve joints of single-layer reticulated shells. J Constr Steel Res 128:601–610
109. Feng R, Wang X, Chen Y et al (2018) Static performance of double-ring joints for freeform single-layer grid shells subjected to a bending moment and shear force. Thin-Walled Struct 131:135–150
110. Oh J, Ju YK, Hwang K et al (2016) FREE joint for a single layer free-form envelope subjected to bending moment. Eng Struct 106:25–35
111. Wang X, Wang X, Wang X et al (2015) Experimental study on stability of aluminum alloy composite structures. J Mechan Eng 48(6):88–90
112. Shi M, Xiang P, Wu M (2018) Experimental investigation on bending and shear performance of two-way aluminum alloy gusset joints. Thin-Walled Struct 122:124–136
113. Shi G, Ban H, Bai Y et al (2013) A novel cast aluminum joint for reticulated shell structures: experimental study and modeling. Adv Struct Eng 16(6):1047–1059
114. Liu H, Chen Z, Xu S et al (2015) Structural behavior of aluminum reticulated shell structures considering semi-rigid and skin effect. Struct Eng Mech 54(1):121–133
115. Weng Z, Zhao Y, Jin Y et al (2018) Classification and development demand of prefabricated joints for spatial grid structures. J Build Struct 39(3):32–38
116. Khalaf AA, Saka MP (2007) Evolutionary structural optimization of steel gusset plates. J Constr Steel Res 63(1):71–81
117. Ren S, Galjaard S (2015) Topology optimisation for steel structural design with additive manufacturing. In: Thomsen M, Tamke M, Gengnagel C, Faircloth B, Scheurer F (eds) Modelling behaviour. Springer, Cham, pp 35–44
118. Seifi H, Xie YM, O'Donnell J et al (2016) Design and fabrication of structural connections using bi-directional evolutionary structural optimization and additive manufacturing. Appl Mech Mater 846:571–576
119. Seifi H, Javan AR, Xu S et al (2018) Design optimization and additive manufacturing of joints in gridshell structures. Eng Struct 160:161–170
120. Chen M (2018) Topology optimization design of spatial structure joints for additive manufacturing. Zhejiang University
121. Yang ZHAO, Minchao CHEN, Zhen WANG (2019) Topology optimization design of cable-stayed structure joints for additive manufacturing. J Archit Struct 40(3):58–68

122. Feng R, Liu F, Zhu J. An assembled joint of outer ring and inner hexagonal space grid structure. CN106836476A, 2017-06-13, Jiangsu
123. Sigmund O, Maute K (2013) Topology optimization approaches. Struct Multidiscip Optim 48(6):1031–1055
124. Bendsøe MP, Kikuchi N (1988) Generating optimal topologies in structural design using a homogenization method. Comput Methods Appl Mech Eng 71(2):197–224
125. Bendsøe MP (1989) Optimal shape design as a material distribution problem. Struct Optim 1(4):193–202
126. Zhou M, Rozvany GIN (1991) The COC algorithm, Part II: topological, geometrical and generalized shape optimization. Comput Methods Appl Mech Eng 89(1–3):309–336
127. Rozvany GIN, Zhou M, Birker T (1992) Generalized shape optimization without homogenization. Struct Optim 4(3–4):250–252
128. Stolpe M, Svanberg K (2001) An alternative interpolation scheme for minimum compliance topology optimization. Struct Multidiscip Optim 22(2):116–124
129. Deaton JD, Grandhi RV (2014) A survey of structural and multidisciplinary continuum topology optimization: post 2000. Struct Multidiscip Optim 49(1):1–38
130. Wang MY, Wang X, Guo D (2003) A level set method for structural topology optimization. Comput Methods Appl Mech Eng 192(1–2):227–246
131. Bourdin B, Chambolle A (2003) Design-dependent loads in topology optimization. In: ESAIM: Control Optim Calcul Variat 9:19–48
132. Xie YM, Steven GP (1993) A simple evolutionary procedure for structural optimization. Comput Struct 49(5):885–896
133. Querin OM, Steven GP, Xie YM (1998) Evolutionary structural optimisation (ESO) using a bidirectional algorithm. Eng Comput 15(8):1031–1048
134. Huang X, Xie M (2010) Evolutionary topology optimization of continuum structures: methods and applications. Wiley
135. National Standard of the People's Republic of China (2008) GB/T 17395–2008 dimension, shape, weight and allowable deviation of seamless steel pipes. Standards Press of China, Beijing
136. Adeli H, Cheng NT (1994) Concurrent genetic algorithms for optimization of large structures. J Aerosp Eng 7(3):276–296

**Open Access** This chapter is licensed under the terms of the Creative Commons Attribution-NonCommercial-NoDerivatives 4.0 International License (http://creativecommons.org/licenses/by-nc-nd/4.0/), which permits any noncommercial use, sharing, distribution and reproduction in any medium or format, as long as you give appropriate credit to the original author(s) and the source, provide a link to the Creative Commons license and indicate if you modified the licensed material. You do not have permission under this license to share adapted material derived from this chapter or parts of it.

The images or other third party material in this chapter are included in the chapter's Creative Commons license, unless indicated otherwise in a credit line to the material. If material is not included in the chapter's Creative Commons license and your intended use is not permitted by statutory regulation or exceeds the permitted use, you will need to obtain permission directly from the copyright holder.

# Chapter 2
# The Stability Mechanism of Single-Layer Gridshells Based on the Theory of Configuration Vulnerability

**Abstract** Stable problem is a special problem of shell structure, and also a controlling factor in the design of single-layer lattice shell structure (Shen and Chen in Stability of reticulated shell structures, Science Press, Beijing, 1999 [1]), that is, the ultimate bearing capacity of single-layer lattice shell structure is generally determined by the stable bearing capacity. Therefore, revealing the instability mechanism of single-layer lattice shell structure is the basis of stable optimization design, which has important theoretical and engineering value for improving the structural safety reserve and excavating the space spanning ability of single-layer lattice shell structure. In 1991, Blockley et al. (Civ Eng Syst 10:301–317, 1993 [2]; Wu et al. in Civ Eng Syst 10:319–333, 1993 [3]) from Bristol University in the UK proposed the vulnerability theory based on joint well-formedness. Based on the joint well-formedness with clear physical meaning, the structural topological hierarchy model is established based on the clustering process to identify the weakest part of the internal connection of the structure; through the declustering process, the various failure modes with vulnerability of the structure are identified. This method has been widely used in the failure analysis of space frame structure (Agarwal et al. in Struct Saf 23:203–220, 2001 [4]), space rod structure (Zhu and Ye in J Eng Mech 140:112–127, 2013 [5]) and cold-formed steel structure (Ye et al. in Appl Sci 7:182, 2017 [6]). The classical configuration vulnerability theory only focuses on the topological configuration of the structure itself, and cannot consider external factors such as load and constraints; at the same time, the theory analyzes the structural characteristics based on the elastic stiffness matrix under the initial configuration, and cannot consider nonlinearity. However, a large number of research results show that the stability of lattice shell structure is closely related to the characteristics of the structure itself, load mode, support constraints and other conditions, and has strong nonlinearity. In this chapter, the joint well-formedness, an important parameter in the theory of configuration vulnerability, is firstly introduced. Then, on the basis of the classical well-formedness, the geometric stiffness matrix is introduced, and a calculation method of joint well-formedness that can consider load conditions, geometric nonlinearity and support constraints is proposed. By analyzing the changes of joint well-formedness before and after the introduction of the geometric stiffness matrix, the instability mechanism of the single-layer lattice shell structure is revealed from the perspective of the loss of stability. Finally, three single-layer lattice shell structures with different scales are taken as examples to verify the instability mechanism of the single-layer lattice shell structure proposed in this chapter.

## 2.1 Joint Well-Formedness of Configuration Vulnerability Theory

Joint conformability is an important parameter in the theory of conformational fragility. At the joint level, conformability is a measure of joint stiffness; at the structure level, conformability reflects the strength of the structure's connection at the joint.

For a structure with $n$ unconstrained joints, its global stiffness matrix **K** can be written in the form of n × n-order block matrix, as shown in Eq. (2.1).

$$\mathbf{K} = \begin{bmatrix} \mathbf{K}_{11} & \cdots & & & \mathbf{K}_{1n} \\ & \ddots & & & \\ \vdots & & \mathbf{K}_{kk} & & \vdots \\ & & & \ddots & \\ \mathbf{K}_{n1} & & \cdots & & \mathbf{K}_{nn} \end{bmatrix} \quad (2.1)$$

In the formula, $\mathbf{K}_{kk}$ is located on the main diagonal of the global stiffness matrix and is a submatrix related to joint $k$. $\mathbf{K}_{kk}$ is a positive definite symmetric matrix, and its dimension C is equal to the number of degrees of freedom of the joint. For space rigid joint joints, C = 6. Then the dimension of the global stiffness matrix is $d = n \times C$.

According to the global stiffness matrix, the global balance equation of the structure is established as shown in Eq. (2.2)

$$\mathbf{F} = \mathbf{KX} \quad (2.2)$$

In the formula (2.2), $\mathbf{F} = \{[\mathbf{F}_1], [\mathbf{F}_2], \ldots, [\mathbf{F}_k], \ldots, [\mathbf{F}_n]\}^T$ is the force vector in the global coordinate system. $\mathbf{X} = \{[\mathbf{X}_1], [\mathbf{X}_2], \ldots, [\mathbf{X}_k], \ldots, [\mathbf{X}_n]\}^T$ is the displacement vector in the global coordinate system.

According to the properties of symmetric positive definite matrix, the correlation stiffness matrix is diagonalized, namely:

$$\mathbf{K} = \mathbf{PHP}^{-1} \quad (2.3)$$

In the formula (2.3), **P** is an orthogonal matrix; **H** is a diagonal matrix, whose main diagonal element is the eigenvalue $\alpha_i$ of **K**, and $\alpha_i$ is greater than 0.

## 2.1 Joint Well-Formedness of Configuration Vulnerability Theory

Substitute Eq. (2.3) into Eq. (2.2) and multiply both sides by the moment $\mathbf{P}^{-1}$, and Eq. (2.4) is obtained.

$$\mathbf{P}^{-1}\mathbf{F} = \mathbf{H}\mathbf{P}^{-1}\mathbf{X} \tag{2.4}$$

Set:

$$\begin{aligned}\mathbf{F}' &= \mathbf{P}^{-1}\mathbf{F} \\ \mathbf{X}' &= \mathbf{P}^{-1}\mathbf{X}\end{aligned} \tag{2.5}$$

Substitute Eq. (2.5) into Eq. (2.4) and we get:

$$\mathbf{F}' = \mathbf{H}\mathbf{X}' \tag{2.6}$$

When $\mathbf{X}'$ is the unit displacement vector, namely $x'_1 = x'_2 = \ldots = x'_d = 1$, Eq. (2.6) can be expanded as follows:

$$\begin{aligned} F'_1 &= \alpha_1 \\ F'_2 &= \alpha_2 \\ &\vdots \\ F'_i &= \alpha_i \\ &\vdots \\ F'_d &= \alpha_d \end{aligned} \tag{2.7}$$

$\alpha_i$ is defined as the principal stiffness coefficient of a joint, and the corresponding eigenvector of $\alpha_i$ is the direction of the corresponding joint's principal displacement axis. According to Eq. (2.7), $\alpha_i$ represents the stiffness of joint $k$ along the direction of the corresponding principal displacement axis, and its size is independent of the selection of the overall coordinate system.

Based on the physical meaning of the principal stiffness coefficient, the well-formedness $q_{k,0}$ [4] of joint $k$ is defined as:

$$q_{k,0} = \det(\mathbf{K_{kk}}) \tag{2.8}$$

According to Eq. (2.8), the well-formedness $q_{k,0}$ of joint $k$ is the product of the stiffness in each principal axis direction, which represents the ability of the joint to resist loads in any direction in the form of a simple scalar. The larger the value is, the greater the overall stiffness of the joint is. Meanwhile, in the structural level, $q_{k,0}$ represents the connection degree of the structure at the joint. The lower the value is, the weaker the structure is at the joint.

## 2.2 Joint Well-Formedness Under Load

Classical vulnerability theory studies the characteristics of the structure itself (structural topological relationship, bar section distribution, etc.), independent of the load, and the analysis process is based on the original configuration, which has the advantage of simple linear analysis. However, it cannot consider the influence of load distribution mode, load amplitude, constraint conditions and nonlinearity on the structure. However, the stability of lattice shell structure is closely related to the above factors.

Under the action of load, the component will produce internal force to balance it. The distribution and amplitude of internal force are closely related to load distribution mode, load amplitude, structure topology, component section, constraints and other factors, which directly affect the stability characteristics of the structure. The unit geometric stiffness matrix is a function of the unit internal force. The overall geometric stiffness matrix integrated by the unit geometric stiffness matrix reflects the distribution of force flow in the structure under the load and constraints. At this time, the tangential stiffness matrix of the structure is the sum of the overall stiffness matrix $\mathbf{K}$ and the geometric stiffness matrix $\mathbf{K}_G$. Therefore, the overall geometric stiffness matrix also reflects the influence of internal force on the overall stiffness.

The overall geometric stiffness matrix is integrated by all the unit geometric stiffness matrices and can be written as follows:

$$\mathbf{K}_G = \sum \mathbf{k}_g^e = \sum \mathbf{k}_{gc}^e + \sum \mathbf{k}_{gt}^e = \mathbf{K}_{GC} + \mathbf{K}_{GT} \tag{2.9}$$

In the equation, $\mathbf{k}_g^e$ is the unit geometric stiffness matrix; $\mathbf{k}_{gc}^e$ is the geometric stiffness matrix of the compression bar; $\mathbf{k}_{gt}^e$ is the geometric stiffness matrix of the tension bar; $\mathbf{K}_{GC}$ is the geometric stiffness matrix integrated by all the compression bars; $\mathbf{K}_{GT}$ is the geometric stiffness matrix integrated by all the tension bars.

Different from traditional frame structures, lattice shells are shape-resistant, whose mechanical characteristics are reflected in the membrane effect, and the stress state of components is mainly axial pressure. The stability of lattice shells is closely related to the distribution of pressure bars and their compressive stress levels. The geometric stiffness matrix $\mathbf{K}_{GC}$ integrated by pressure bars can directly reflect the weakening of structural stiffness by axial pressure, and reflect the instability trend of the structure. Therefore, the well-formedness $q_{k,1}$ of joint $k$ under load is defined as follows:

$$q_{k,1} = \det(\mathbf{K}_{kk} + \mathbf{K}_{GCkk}) \tag{2.10}$$

In the equation, $K_{kk}$ is shown in Eq. (2.1); $\mathbf{K}_{GCkk}$ is the sub-stiffness matrix related to joint $k$ in $\mathbf{K}_{GC}$; $\mathbf{K}_{GC}$ is shown in Eq. (2.9).

$q_{k,1}$ represents the overall stiffness of joint $k$ after considering external factors. By comparison with $q_{k,0}$, $q_{k,1}$ is the ability of joint to resist loads in all directions after considering the weakening of the original structure stiffness by compressive stress.

## 2.3 Relative Gradient of Joint Well-Formedness

$Q_{k,0}$ is the initial well-formedness of joint $k$, indicating the initial stiffness of the joint, which is only related to the structure itself. $q_{k,1}$ is the well-formedness of the joint under the load, which is the stiffness of the joint after considering the influence of external factors. Under the load, the lattice shell structure will lose its ability to resist the load due to the stiffness softening caused by compressive stress. By examining the changes in the well-formedness of the joint before and after the load, the relative change gradient $gra\_r_k$ of the well-formedness of the joint $k$ is defined, as shown in Eq. (2.11), to quantitatively measure the degree of joint softening.

$$gra\_r_k = \frac{q_{k,1} - q_{k,0}}{q_{k,0}} \qquad (2.11)$$

In the equation, $q_{k,0}$ is shown in Eq. (2.8); $q_{k,1}$ is shown in Eq. (2.10).

$gra\_r_k$ is a reference to the configuration of the joint itself to measure the extent of degradation of joint stiffness by external factors. Therefore, $gra\_r_k$ comprehensively measures the stability characteristics of joint $k$ from both internal and external factors of the structure. Since $q_{k,1}$ is generally less than $q_{k,0}$, $gra\_r_k$ is negative. The lower the value is, the more significant the degradation of joint stiffness is. According to the value and distribution of $gra\_r_k$, the instability area can be identified and the key joint that determines the structural stability bearing capacity can be accurately identified.

## 2.4 Verification of Stability Tracking Strategy Using Arc Length Method

The arc length method nonlinear tracking can successfully cross the stable critical point, solve the stiffness matrix singularity near the critical point and the non-convergence problem, not only can calculate the stable bearing capacity of lattice shell structure, but also can track the post-buckling behavior of the structure, realize the whole process of tracking the balance path. It is currently recognized as the calculation method for solving the stable bearing capacity of the structure. The correctness of the research results of the extended configuration vulnerability theory is proposed to be verified by the results obtained by the arc length method stability tracking.

As shown in Fig. 2.1, the space rigid frame is supported by sliding hinges around. Through the comparative analysis of the load displacement curve in Fig. 2.1, it can be seen that the critical load obtained in this paper is slightly lower than the result obtained by Meek and Tan [7] using the arc length method, and is close to the load displacement curve obtained by Papadrakakis [8] using the two-vector iteration method, which verifies the correctness of the solving strategy of the arc length method in this paper.

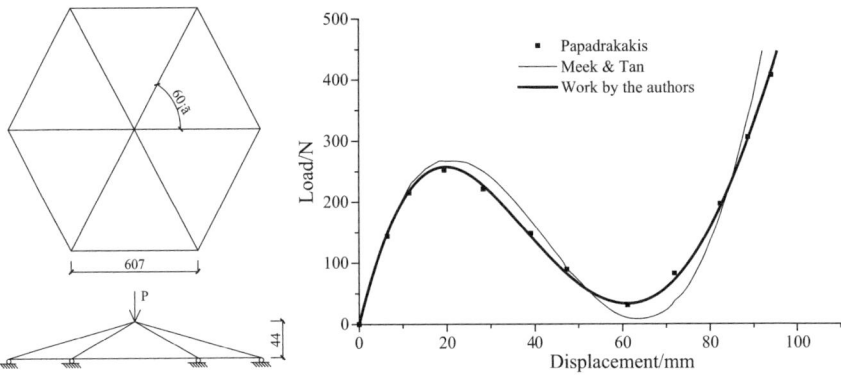

**Fig. 2.1** Space rigid frame and load displacement curve

## 2.5 Example 1: Single-Layer Reticulated Shell with 14 m Span

As shown in Fig. 2.2, the K6 single-layer spherical reticulated shell has a span of 14 m, a height of 2.5 m, fixed supports around, and circular steel pipes with cross-section of $\Phi 114 \times 8$ and elastic modulus $E = 2.06 \times 10^{11}$ N/m². The well-formedness analysis and arc-length method elastic stability analysis are conducted on the structure under full span uniform load, half span uniform load and vertex concentrated load, respectively.

### 2.5.1 Full Span Uniform Load Mode

Apply vertical downward load F = 1 N on all joints, and the load amplitude $\sum F = 19$ N. According to Eq. (2.11), the relative change gradient $gra\_r$ of well-formedness of each joint can be obtained, as shown in Table 2.1. On the lattice shell structure, draw the distribution cloud map of joint $gra\_r$, as shown in Fig. 2.3. In Fig. 2.3, the three-dimensional coordinates of joints are the actual spatial position of joints, the color of joints is determined by the value of $gra\_r$ of the joint, and the line segments represent the bars connecting joints.

According to Table 2.1 and Fig. 2.3, $gra\_r$ of all joints is similar, indicating that the stiffness degradation degree of all joints is similar and the structure is able to resist load as a whole; among all joints, $gra\_r$ of the second circle joint is the lowest (denoted as $gra\_r_{\min}$), $gra\_r_{\min} = -1.294 \times 10^{-6}$, indicating that the stiffness degradation of these joints is the most significant.

The arc length method is used to conduct elastic stability tracking of the single-layer reticulated shell as shown in Fig. 2.2 under full span uniform load. The stable critical load proportion coefficient $\lambda_{cr} = 1.07578 \times 10^6$, the stable critical load

## 2.5 Example 1: Single-Layer Reticulated Shell with 14 m Span

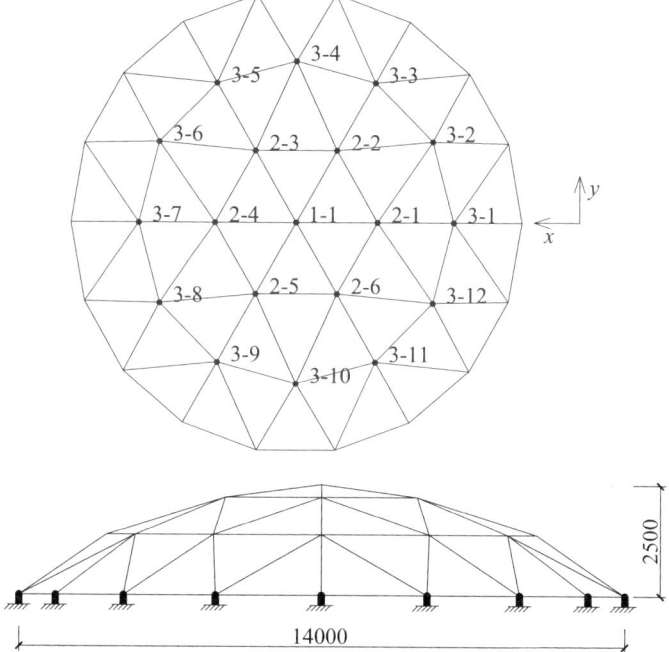

**Fig. 2.2** 14 m span single layer spherical reticulated shell and joint number (unit: mm)

$P_{cr} = \lambda_{cr} \times F = 1.07578 \times 10^6$ N/joint, and the structural stable bearing capacity $\sum P_{cr} = \lambda_{cr} \times \sum F = 2.04398 \times 10^7$ N. The structural instability mode is overall instability, as shown in Fig. 2.4. When instability occurs, the vertical deflection of each joint is shown in Table 2.2, where the vertical deflection of the second circle joint is greater than that of other joints.

According to Figs. 2.3 and 2.4, when $gra\_r$ of all joints of the structure is similar, the structure is fully softened and the structural instability mode is overall instability. According to Tables 2.1 and 2.2, the joint stiffness degradation corresponding to $gra\_r_{min}$ is the most significant, and the joint will experience larger deformation during instability.

### 2.5.2 *Half Span Uniform Load Mode*

The load is distributed in the area where $Y \geq 0$ in Fig. 2.2, that is, 1 N vertical load is applied to joints 1–1, 2–1, 3–1, 2–4, and 3–7, and 2 N vertical load is applied to joints 2–2, 2–3, 3–2, 3–3, 3–4, 3–5, and 3–6. The load amplitude $\sum F = 19$ N, and the total load is consistent with the full span uniform load mode. The relative change gradient of well-formedness of each joint can be obtained by Eq. (2.11), as

**Table 2.1** Each joint's *gra_r* of 14 m span single layer lattice shell under different load modes

| Joint number | Full span uniform load $gra\_r/\times 10^{-6}$ | Vertex concentrated load $gra\_r/\times 10^{-6}$ | Half span uniform load $gra\_r/\times 10^{-6}$ |
|---|---|---|---|
| 1–1 | − 1.278 | − 1.278 | − 19.689 |
| 2–1 | − 1.294 | − 1.324 | − 5.153 |
| 2–2 | − 1.294 | − 2.566 | − 5.153 |
| 2–3 | − 1.294 | − 2.566 | − 5.153 |
| 2–4 | − 1.294 | − 1.324 | − 5.153 |
| 2–5 | − 1.294 | − 0.576 | − 5.153 |
| 2–6 | − 1.294 | − 0.576 | − 5.153 |
| 3–1 | − 1.161 | − 1.438 | − 1.518 |
| 3–2 | − 1.192 | − 2.345 | − 1.357 |
| 3–3 | − 1.161 | − 2.325 | − 1.518 |
| 3–4 | − 1.192 | − 2.401 | − 1.357 |
| 3–5 | − 1.161 | − 2.325 | − 1.518 |
| 3–6 | − 1.192 | − 2.345 | − 1.357 |
| 3–7 | − 1.161 | − 1.438 | − 1.518 |
| 3–8 | − 1.192 | − 0.684 | − 1.357 |
| 3–9 | − 1.161 | − 0.416 | − 1.518 |
| 3–10 | − 1.192 | − 0.444 | − 1.357 |
| 3–11 | − 1.161 | − 0.416 | − 1.518 |
| 3–12 | − 1.192 | − 0.684 | − 1.357 |

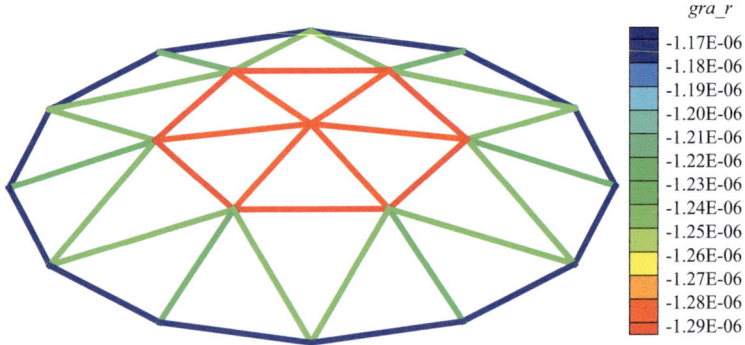

**Fig. 2.3** *gra_r* distribution of joints in 14 m span single-layer lattice shell structure under full span uniform load

## 2.5 Example 1: Single-Layer Reticulated Shell with 14 m Span

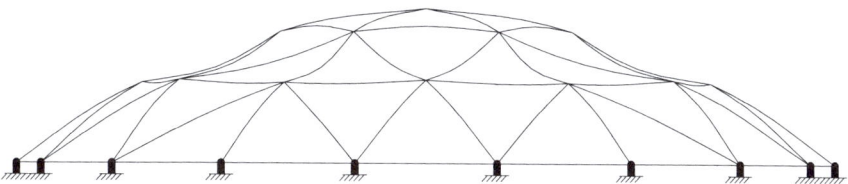

**Fig. 2.4** Overall instability mode of 14 m-span reticulated shell under full span uniformly distributed load

**Table 2.2** Joint deflection of 14 m-span lattice shell when it fails under full span uniformly distributed load

| Joint number | 1–1 | 2–1 | 2–2 | 3–1 | 3–2 | 3–3 |
|---|---|---|---|---|---|---|
| Displacement/$10^{-2}$ m | 3.3983 | **7.1879** | **7.1879** | 4.7372 | 2.6619 | 4.7372 |

shown in Table 2.1. On the lattice shell structure, the distribution cloud map of joint *gra_r* is drawn, as shown in Fig. 2.5. In Fig. 2.5, the three-dimensional coordinates of joints are the actual spatial position of joints, the color of joints is determined by the *gra_r* value of the joint, and the line segments represent the bars connecting joints. Figure 2.5 shows that: the *gra_r* value of joints in the load-bearing area is low, shown in red in Fig. 2.5; the *gra_r* value of joints in the non-load-bearing area is high, shown in blue in Fig. 2.5.

Based on Table 2.1 and Fig. 2.5, it can be concluded that *gra_r* of joints in the load-bearing area is significantly lower than that of joints in the non-load-bearing area, indicating that the degree of stiffness degradation of joints in the load-bearing area is significantly greater than that of joints in the non-load-bearing area; *gra_r* of joints 2–2 and 2–3 is the smallest, $gra\_r_{\min} = -2.566 \times 10^{-6}$, indicating that the degree of stiffness degradation of these two joints is the most significant.

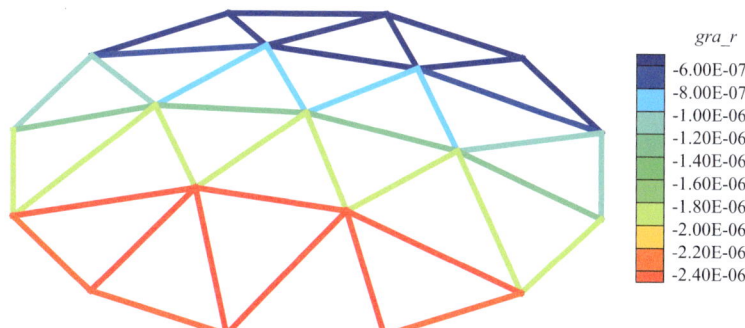

**Fig. 2.5** *gra_r* distribution of joints in 14 m span single-layer lattice shell structure under uniformly distributed load at half span

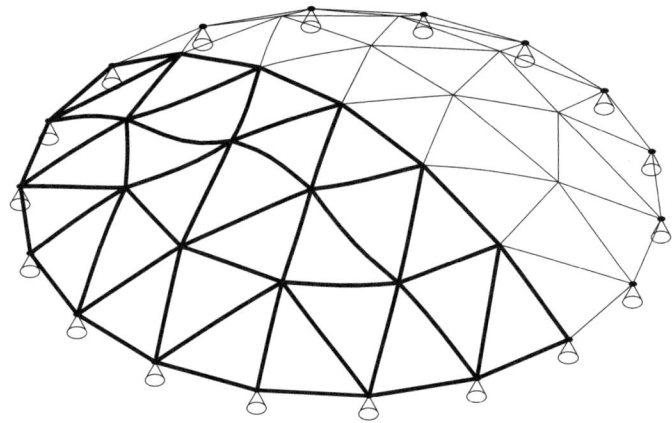

**Fig. 2.6** Local instability mode of 14 m span grid shell under half span uniformly distributed load

**Table 2.3** Characteristics of partial joints when 14 m span lattice shell fails under half-span uniformly distributed load

| Joint | 2–2 | 3–3 | 2–5 | 3–10 |
|---|---|---|---|---|
| $gra\_r/\times 10^{-6}$ | 2.566 | 2.325 | 0.576 | 0.444 |
| Displacement/$\times 10^{-1}$ m | 1.0977 | 0.5235 | 0.3142 | 0.2052 |
| Rotation/$\times 10^{-2}$ rad | 11.856 | 7.4953 | 0.3439 | 0.9209 |

*Note* Displacement $= (u_x^2 + u_y^2 + u_z^2)^{1/2}$, Rotation $= (rot_x^2 + rot_y^2 + rot_z^2)^{1/2}$

The arc length method is adopted to track the elastic stability of the single-layer reticulated shell as shown in Fig. 2.2 under the uniform load of half span. The stable critical load proportion coefficient $\lambda_{cr} = 4.25161 \times 10^5$, and the structural stable bearing capacity $\sum P_{cr} = \lambda_{cr} \times \sum F = 8.07806 \times 10^6$ N are obtained through calculation. The structural instability mode is local instability, as shown in Fig. 2.6. The representative joint displacement during instability is shown in Table 2.3.

According to Figs. 2.5 and 2.6, when $gra\_r$ of some joints is significantly lower than other joints, the structure becomes partially softened and the instability mode is a local instability mode in the softening area. According to Tables 2.1 and 2.3, the joint stiffness degradation corresponding to $gra\_r_{min}$ is the most significant and the joint becomes unstable first in the loading process.

### 2.5.3 Vertex Concentrated Load Mode

Vertical downward load of 19 N is applied at vertex 1–1 (Fig. 2.2), the load amplitude $\sum F = 19$ N, and the total load is consistent with the full span uniform load mode. The relative change gradient of well-formedness of each joint can be obtained by

Eq. (2.11), see Table 2.1. On the lattice shell structure, the distribution cloud map of joint gra_r is drawn, as shown in Fig. 2.7. In Fig. 2.7, the three-dimensional coordinates of joints are the actual spatial position of joints, the color of joints is determined by the value of gra_r of the joint, and the line segments represent the bars connecting joints.

Combining Table 2.1 and Fig. 2.7, it can be concluded that gra_r of load-bearing joints is significantly lower than that of other joints, $gra\_r_{min} = -19.689 \times 10^{-6}$, indicating that the degree of stiffness degradation of load-bearing joints is significantly higher than that of other joints by an order of magnitude.

The arc length method is adopted to track the elastic stability of the single-layer reticulated shell as shown in Fig. 2.2 under the vertex concentrated load. The stable critical load proportional coefficient $\lambda_{cr} = 4.31580 \times 10^4$ and the structural stable bearing capacity $\sum P_{cr} = \lambda_{cr} \times \sum F = 8.20002 \times 10^5$ N are obtained through calculation. The structural instability mode is point instability, as shown in Fig. 2.8, and the displacement of some joints is shown in Table 2.4. During the instability, the deflection of the vertex is two orders of magnitude greater than that of the surrounding joints, and the angular displacement of the second layer joint is obviously greater than that of other joints.

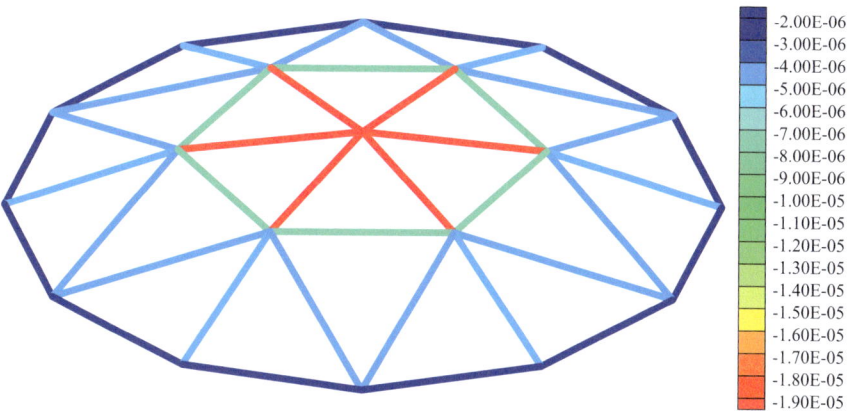

**Fig. 2.7** Nodal gra_r distribution of 14 m span single-layer lattice shell structure under vertex-concentrated load

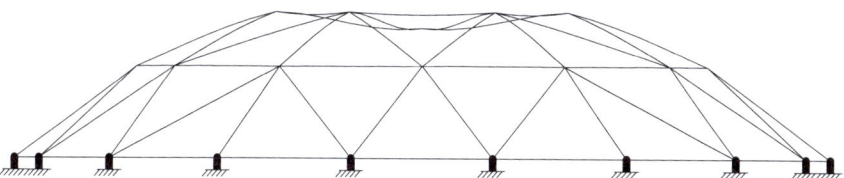

**Fig. 2.8** Point instability mode of 14 m span lattice shell under concentrated load at the vertex

**Table 2.4** Displacement of partial joints when 14 m span lattice shell fails under concentrated load at vertex

| Joint | 1–1 | 2–1 | 3–1 |
|---|---|---|---|
| Displacement/×$10^{-2}$ m | **14.3570** | − 0.3865 | − 0.4124 |
| Rotation/×$10^{-2}$ rad | 2.7077 × $10^{-12}$ | **2.8041** | 0.6790 |

Figures 2.7 and 2.8 indicate that when the *gra_r* of individual joints is significantly lower than that of other joints, the individual joints in the structure will soften significantly, and the structure will show point instability mode during stable loading. Tables 2.1 and 2.4 indicate that the *gra_r* of the vertex is the lowest, and the vertex will fail first during stable loading; the *gra_r* of the second circle joint is the second lowest, significantly lower than that of the third circle joint, and the second circle joint will experience a large angular displacement during instability.

## *2.5.4 Structure Instability Mechanism and Most Unfavorable Load Mode*

The characteristics of well-formedness change and elastic stability of the single-layer lattice shell structure shown in Fig. 2.2 under three load modes were studied, and the instability modes presented included global instability, local instability and point instability. No matter what kind of instability mode, the development of instability region always starts from the joint with the most significant softening degree (i.e. the joint corresponding to $gra\_r_{min}$), and its development and propagation are consistent with the distribution law of *gra_r*. Therefore, under a given load mode, the joint with the most significant well-formedness degradation is the key joint that affects the structural stability performance, and is closely related to the critical load of structural stability. See Table 2.5 for $gra\_r_{min}$ and $\Sigma P_{cr}$ under the above three load modes. It is obvious that the vertex-concentrated load mode is the most unfavorable stable load mode.

**Table 2.5** The most unfavorable stable load mode of lattice shell structure with 14 m span

| Load mode | $gra\_r_{min}/\times 10^{-6}$ | $\Sigma P_{cr}/\times 10^5$ N |
|---|---|---|
| Vertex concentrated load mode | − 19.689 | 8.20002 |
| Half span uniform load mode | − 2.566 | 80.7806 |
| Full span uniform load mode | − 1.294 | 204.398 |

## 2.6 Example 2: Single-Layer Reticulated Shell with 22 m Span

As shown in Fig. 2.9, the K6 single-layer lattice shell has a span of 22 m, a height of 11 m, fixed supports around, and the bar section is $\Phi 114 \times 8$ round steel pipe with an elastic modulus of $E = 2.06 \times 1011$ N/m². The steel amount used for the bar is 3.041 m³. The structural deformation variation characteristics and structural elastic stability are analyzed respectively under the full span uniform load mode, half span uniform load mode and vertex concentrated load mode.

### 2.6.1 Full Span Uniform Load Mode

Each non-constrained joint is subjected to a vertical downward load of 1 N, and the load amplitude $\sum F = 127$ N. According to Eq. (2.11), the relative change gradient of well-formedness of each joint can be obtained. The projected position of each free joint to the horizontal plane is taken as the X and Y coordinates, and the *gra_r* of joints is taken as the Z coordinate to draw the spatial distribution map of joint *gra_r* distribution. In order to enhance the expression effect, the *gra_r* of joints is linearly interpolated between discrete joints to generate the spatial distribution cloud map, as shown in Fig. 2.10 The line segment in Fig. 2.10 represents the rod connecting the joints, which can reflect the structural topological connection relationship.

Figure 2.10 indicates that under the uniformly distributed full-span load, all joints exhibit varying degrees of configuration degradation. The softening degree of the top joints is more significant than that of the bottom joints, and the *gra_r* of the second ring (from top to bottom, with the vertex as the first ring) is lowest, with $gra\_r_{min} = -1.267 \times 10^{-6}$.

By using the arc length method to perform elastic stability tracking on the single-layer shell structure shown in Fig. 2.9 under full-span uniformly distributed load, we

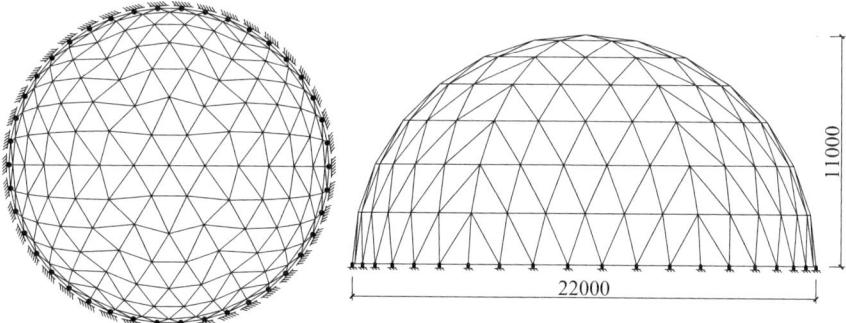

**Fig. 2.9** 22 m span single layer spherical lattice shell (unit: mm)

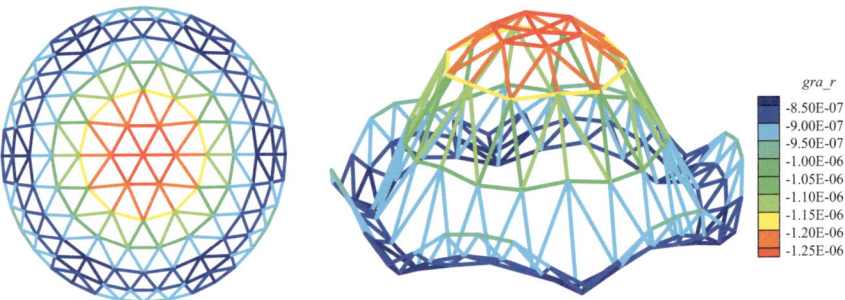

**Fig. 2.10** *gra_r* distribution of joints in 22 m span single-layer lattice shell structure under full span uniform load

can obtain the critical load proportion coefficient $\lambda_{cr} = 8.3438 \times 10^5$. Therefore, the stable critical load $P_{cr} = \lambda_{cr} \times F = 8.3438 \times 10^5$ N/joint. The overall structural stability bearing capacity $\sum P_{cr} = \lambda_{cr} \times \sum F = 1.05966 \times 10^8$ N. The instability mode of the structure exhibits global instability, as shown in Fig. 2.11a, and the vertical displacement distribution cloud map is shown in Fig. 2.11b.

Combining Figs. 2.10 and 2.11, it can be observed that under the full-span uniformly distributed load, the *gra_r* of all joints in the structure is similar, indicating a comprehensive softening of the structure. During stable loading, the structural instability mode exhibits overall instability. The areas with smaller *gra_r* in Fig. 2.10 (i.e., the red area) experience significant instability deformation in Fig. 2.11b (corresponding to the red area in the figure), while areas with larger *gra_r* in Fig. 2.10 (i.e., the blue area) exhibit less unstable deformation in Fig. 2.11b (corresponding to the blue area).

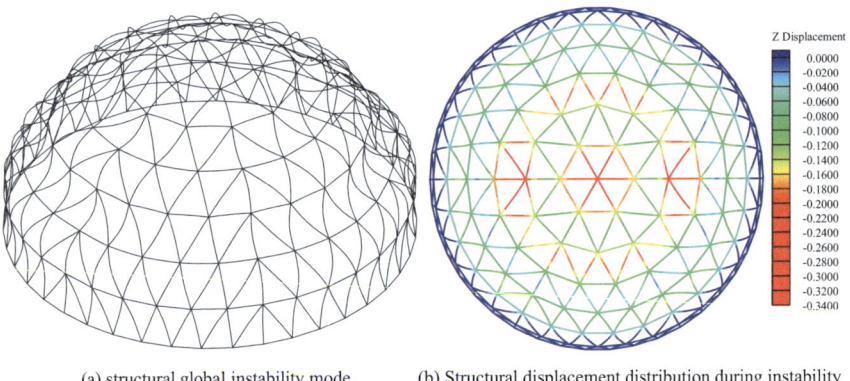

(a) structural global instability mode    (b) Structural displacement distribution during instability

**Fig. 2.11** Overall instability mode of 22 m-span reticulated shell under full span uniformly distributed load (unit: m)

## 2.6.2 Half Span Uniform Load Mode

Apply the half-span uniform load to the K6 single-layer reticulated shell as shown in Fig. 2.9 in the same way as in Sect. 2.5.2. Keep the load amplitude $\sum F = 127$ N the same as that of full-span uniform load mode. The relative change gradient of well-formedness of each joint can be obtained from Eq. (2.11). Draw the spatial distribution map of joint $gra\_r$ distribution by taking the projected position of each free joint to the horizontal plane as X, Y coordinates and joint $gra\_r$ as Z coordinates. In order to enhance the expression effect, linear interpolation of joint $gra\_r$ is performed between discrete joints to generate the spatial distribution cloud map, as shown in Fig. 2.12. The line segments in Fig. 2.12 represent the bars connecting the joints, which can reflect the structural topological connection relationship. The value of joint $gra\_r$ in load-bearing area is significantly lower than that of joint in non-load-bearing area, which is red in Fig. 2.12, indicating that the stiffness degradation degree of joint in load-bearing area is significantly greater than that of joint in non-load-bearing area. The joint with the lowest $gra\_r$ is located in the second circle of load-bearing area, as shown in the bold joint in Fig. 2.12, $gra\_r_{min} = -2.496 \times 10^{-6}$.

The arc length method is used to track the elastic stability of the single-layer reticulated shell as shown in Fig. 2.9 under the uniform load of half span. The stable critical load proportion coefficient $\lambda_{cr} = 3.9829 \times 10^5$, and the structural stable bearing capacity $\sum P_{cr} = \lambda_{cr} \times \sum F = 5.05828 \times 10^7$ N are obtained by calculation. The structural instability mode is local instability, as shown in Fig. 2.13a, where the black bold part is the loading area. When the structure is instable, the vertical displacement nephogram is shown in Fig. 2.13b.

Combining Figs. 2.12 and 2.13, it can be observed that under the half-span uniform load, some joints exhibit significantly lower $gra\_r$ values compared to others, indicating partial softening of the structure. When instability occurs, it exhibits a local softening zone instability mode. The red areas in Fig. 2.12 with significant softening

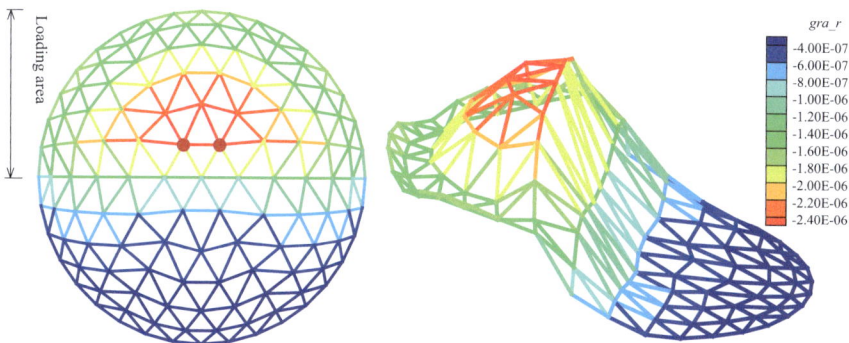

**Fig. 2.12** $gra\_r$ distribution of 22 m span single-layer lattice shell structure under uniformly distributed load at half span

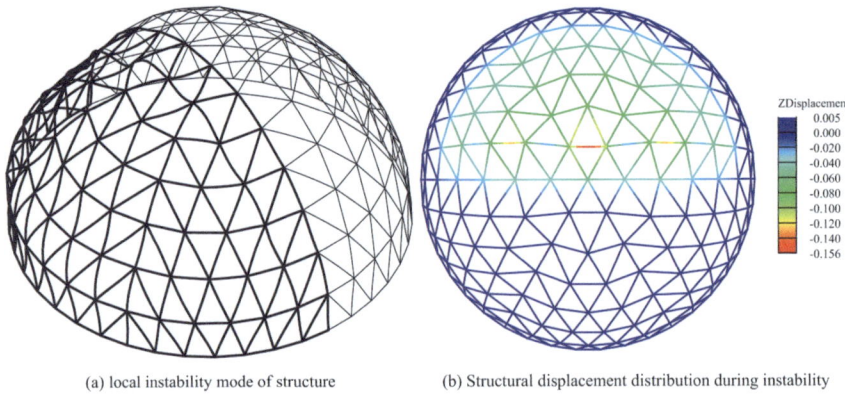

(a) local instability mode of structure    (b) Structural displacement distribution during instability

**Fig. 2.13** Local instability mode of 22 m span grid shell under half span uniformly distributed load (unit: m)

correspond to significant unstable deformations in Fig. 2.13b (corresponding to the red areas). The joints with the most significant degree of softening (i.e., corresponding to $gra\_r_{min}$) also show the most significant deformations in Fig. 2.13b. Joints outside of the loading area have higher $gra\_r$ values, indicating less noticeable softening compared to the loaded area. These joints remain stable during the stable loading process; hence their deformation is smaller in Fig. 2.13b (corresponding to blue areas).

### 2.6.3 Vertex Concentrated Load Mode

Vertical load of 127 N is applied at the vertex, and the load amplitude $\sum F = 127$ N, and the total load is consistent with the full span uniform load mode. According to Eq. (2.11), the relative change gradient of well-formedness of each joint can be obtained. The position of each free joint projected to the horizontal plane is taken as the X and Y coordinates, and the $gra\_r$ of joints is taken as the Z coordinate to draw the spatial distribution map of joint $gra\_r$ distribution. In order to enhance the expression effect, the $gra\_r$ of joints is linearly interpolated between discrete joints to generate the spatial distribution cloud map, as shown in Fig. 2.14. The line segment in Fig. 2.14 represents the bar connecting the joints, which can reflect the structural topological connection relationship. According to Fig. 2.14, the $gra\_r$ of load-bearing joint is significantly lower than that of other joints, indicating that the stiffness degradation of load-bearing joint is the most significant, and its $gra\_r_{min} = -1.274 \times 10^{-4}$.

Using the arc length method, the elastic stability tracking of the single-layer grid shell shown in Fig. 2.9 under concentrated vertex load was conducted. The calculation yielded a critical stability load factor $\lambda_{cr} = 6.389 \times 10^3$, and the structural stable

## 2.6 Example 2: Single-Layer Reticulated Shell with 22 m Span

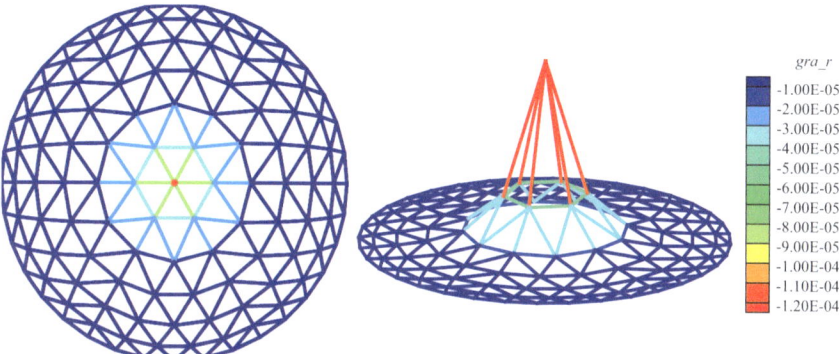

**Fig. 2.14** Nodal *gra_r* distribution of 22 m span single-layer lattice shell structure under vertex-concentrated load

bearing capacity $\sum P_{cr} = \lambda_{cr} \times \sum F = 8.11403 \times 10^5$ N. The structural instability mode exhibits point instability, as shown in Fig. 2.15a, and the vertical displacement cloud map during instability is depicted in Fig. 2.15b.

Combining Figs. 2.14 and 2.15, it can be concluded that when the *gra_r* of individual joints is significantly lower than other joints, it indicates significant softening of these joints, and the structure exhibits a point instability mode during stable loading process. The joint with the most significant softening in Fig. 2.14 experiences significant instability deformation in Fig. 2.15b. Joints in non-loaded areas have higher *gra_r* values, indicating that the degree of softening in this area is less pronounced compared to loaded joints. These non-loaded areas remain stable during the stable loading process; therefore, they exhibit smaller deformations in Fig. 2.15b (corresponding to the blue area).

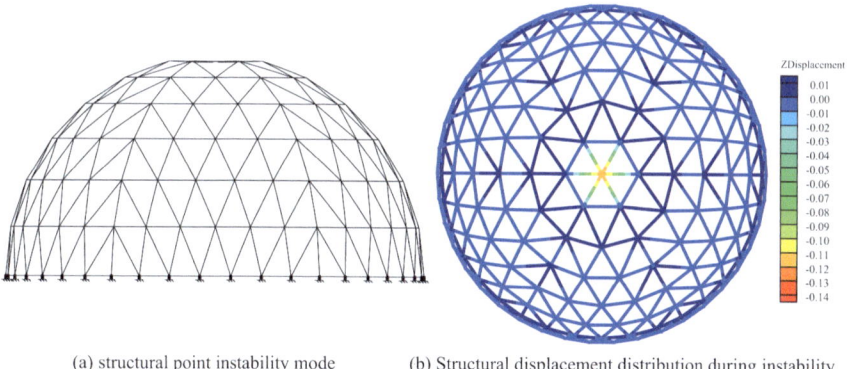

(a) structural point instability mode    (b) Structural displacement distribution during instability

**Fig. 2.15** Point Instability mode of 22 m span lattice shell under concentrated load at the vertex (unit: m)

## 2.6.4 Structure Instability Mechanism and Most Unfavorable Load Mode

The characteristics of the deformation degree and elastic stability of the small-span single-layer lattice shell structure shown in Fig. 2.9 under three load modes were studied, and the instability modes presented included global instability, local instability and point instability. No matter what kind of instability mode, the development of the instability region always starts from the joint with the most significant softening degree (i.e. the joint corresponding to $gra\_r_{min}$), and its development and propagation are consistent with the distribution law of $gra\_r$. Therefore, under a given load mode, the joint with the most significant deformation degree is the key joint that affects the structural stability performance, and is closely related to the critical load of structural stability. See Table 2.6 for $gra\_r_{min}$ and $\Sigma P_{cr}$ under the above three load modes. It can be seen from Table 2.6 that the lower the $gra\_r_{min}$ is, the lower the structural stability bearing capacity is for the same structure under different load modes. It is obvious that the vertex concentrated load is the most unfavorable load mode.

## 2.7 Example 3: Single-Layer Reticulated Shell with 50 m Span

As shown in Fig. 2.16, the K6 single-layer lattice shell has a span of 50 m, a height of 20 m, fixed supports around, and the bar section is $\Phi 114 \times 8$ round steel pipe, with an elastic modulus $E = 2.06 \times 10^{11}$ N/m². The total steel consumption of the bar is 9.738 m³. This structure is a medium span single-layer lattice shell structure. The structural configuration variation characteristics and structural elastic stability are analyzed respectively under the full span uniform load, half span uniform load and vertex concentrated load modes.

**Table 2.6** The most unfavorable stable load mode of lattice shell structure with 22 m span

| Load mode | $gra\_r_{min}/\times 10^{-6}$ | $\Sigma P_{cr}/\times 10^5$ N |
|---|---|---|
| Vertex concentrated load mode | − 127.4 | 8.11403 |
| Half span uniform load mode | − 2.496 | 505.828 |
| Full span uniform load mode | − 1.267 | 1059.66 |

## 2.7 Example 3: Single-Layer Reticulated Shell with 50 m Span

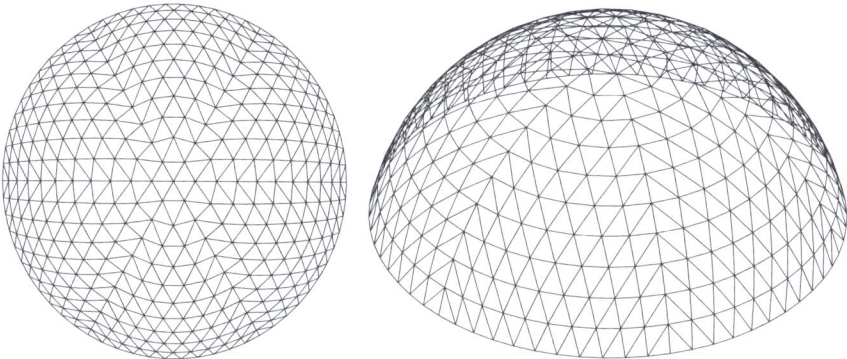

**Fig. 2.16** 50 m span single-layer lattice shell

### 2.7.1 Full Span Uniform Load Mode

Each free joint is subjected to a vertical downward load of 1 N, and the load amplitude $\sum F = 331$ N. According to Eq. (2.11), the relative change gradient of well-formedness of each joint can be obtained. The projected position of each free joint to the horizontal plane is taken as the X and Y coordinates, and the $gra\_r$ of joints is taken as the Z coordinate to draw the spatial distribution map of joint $gra\_r$ distribution. In order to enhance the expression effect, the $gra\_r$ of joints is linearly interpolated between discrete joints to generate the spatial distribution cloud map, as shown in Fig. 2.17. The line segment in Fig. 2.17 represents the rod connecting the joints, which can reflect the structural topological connection relationship.

Figure 2.17 indicates that under the uniformly distributed full-span load, all joints' $gra\_r$ are in the same order of magnitude, indicating that the softening degree of each joint is similar. The softening degree of top-level joints is significantly higher than

**Fig. 2.17** $gra\_r$ distribution of joints in 50 m span single-layer lattice shell structure under full span uniform load

that of bottom-level joints, and the *gra_r* of joints in the third ring (from top to bottom, with the vertex as the first ring) is lowest, with $gra\_r_{min} = -4.304 \times 10^{-6}$.

Using the arc length method to perform elastic stability tracking on the single-layer shell structure shown in Fig. 2.16 under full-span uniformly distributed load, we obtain a critical load proportion coefficient of $\lambda_{cr} = 2.6732 \times 10^5$. Therefore, the stable critical load $P_{cr} = \lambda_{cr} \times F = 2.6732 \times 10^5$ N/Joint, and the overall structural stability bearing capacity $\sum P_{cr} = \lambda_{cr} \times \sum F = 8.8483 \times 10^7$ N. The structural instability mode exhibits overall instability, as shown in Fig. 2.18. By projecting each joint onto the horizontal plane with its position as X and Y coordinates and taking the vertical displacement of joints at instability as Z coordinates, we can obtain a cloud map of vertical displacement distribution for the structure during instability, as shown in Fig. 2.19. In particular, Fig. 2.19a depicts a significantly deformed three-dimensional deformation map of the top region while Fig. 2.19b shows a planar cloud map of overall structural vertical displacement.

Combining Figs. 2.17 and 2.18, we can observe that under the full-span uniformly distributed load, the *gra_r* of all joints in the structure are similar, indicating a comprehensive softening of the structure. The top five circle joints with smaller *gra_r* in Fig. 2.17 (i.e. the red area in the figure) have significant deformation in Fig. 2.19 (corresponding to the red area in the figure). The joint corresponding to $gra\_r_{min}$ is the third circle joint, which has the most significant deformation in Fig. 2.19a; the

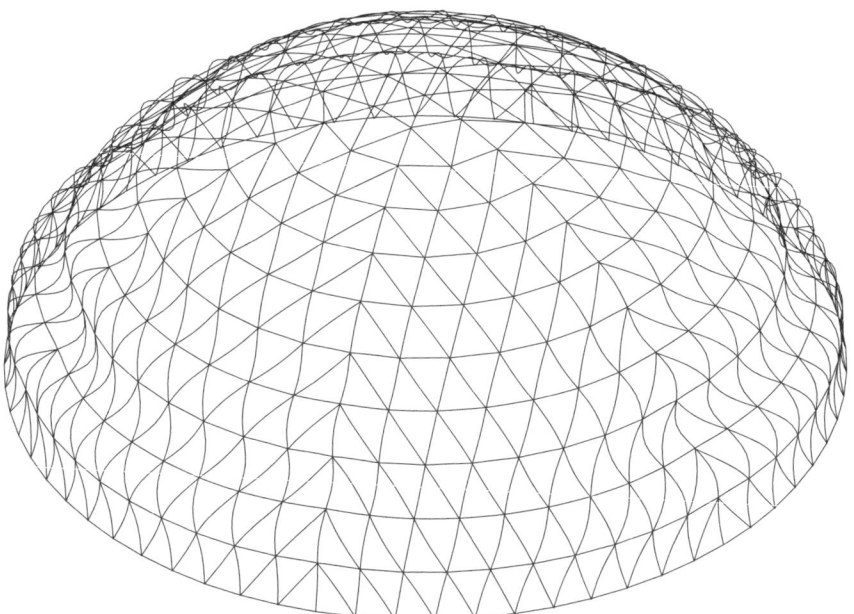

**Fig. 2.18** Overall instability mode of 50 m-span reticulated shell under full span uniformly distributed load

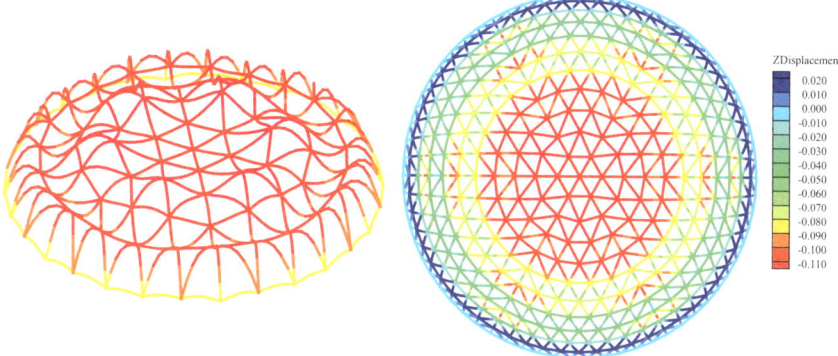

(a) 3D Schematic of Vertical Displacement in Significant Area of Top Deformation  (b) 2D Schematic Diagram of Overall Vertical Displacement of Structure

**Fig. 2.19** Vertical displacement of 50 m-span lattice shell when it fails under full span uniform load (Unit: m)

area with larger gra_r in Fig. 2.17 (i.e. the blue area in the figure) has small instability deformation in Fig. 2.19b (corresponding to the blue area in the figure).

### 2.7.2 Half Span Uniform Load Mode

Apply the half-span uniform load to the medium span K6 single-layer reticulated shell as shown in Fig. 2.16 in the same way as in Sect. 2.5.2. Keep the load amplitude $\sum F = 331$ N the same as that of full-span uniform load mode. The relative change gradient of well-formedness of each joint can be obtained from Eq. (2.11). Draw the spatial distribution map of joint $gra\_r$ distribution by taking the projected position of each free joint to the horizontal plane as X, Y coordinates and joint $gra\_r$ as Z coordinates. In order to enhance the expression effect, linear interpolation of joint $gra\_r$ is performed between discrete joints to generate spatial distribution cloud map, as shown in Fig. 2.20. The line segment in Fig. 2.20 represents the bar connecting the joints, which can reflect the structural topological connection relationship. It can be obtained from Fig. 2.20 that the value of $gra\_r$ of joints in load-bearing area is significantly lower than that of joints in non-load-bearing area, which is red in Fig. 2.20, indicating that the degree of stiffness degradation of joints in load-bearing area is significantly greater than that of joints in non-load-bearing area. The joint with the lowest gra_r is located in the third circle of load-bearing area, as shown in the bold joint in Fig. 2.20, $gra\_r_{\min} = -8.731 \times 10^{-6}$.

The arc length method is used to track the elastic stability of the single-layer reticulated shell shown in Fig. 2.16 under the uniform load of half span. The stable critical load proportional coefficient $\lambda_{cr} = 1.1804 \times 10^5$, and the structural stable bearing capacity $\sum P_{cr} = \lambda_{cr} \times \sum F = 3.9071 \times 10^7$ N are obtained by calculation.

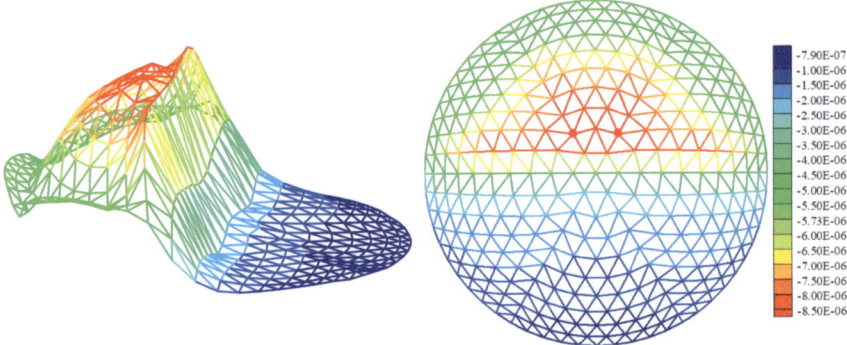

**Fig. 2.20** *gra_r* distribution of joints in 50 m span single-layer lattice shell under uniformly distributed load at half span

The structural instability mode is local instability, as shown in Fig. 2.21a, where the black bold part is the loading area. When the structure is instable, the vertical displacement distribution is shown in Fig. 2.21b.

According to Figs. 2.20 and 2.21, under the uniformly distributed load of half span, *gra_r* of some joints is significantly lower than that of other joints, and the structure is partially softened, and the instability is presented as local instability mode in the softened area; the red area with significant softening degree in Fig. 2.20 has significant instability deformation in Fig. 2.21b, and the deformation of the joint area with the most significant softening degree (i.e. the peripheral joints corresponding to $gra\_r_{min}$) is also the most significant in Fig. 2.21; the joint *gra_r* in the non-loading area is high, indicating that the softening degree of this area is not as significant as that of the loaded area, and it remains stable in the process of stable loading, so the deformation of this area is small in Fig. 2.21b (corresponding to the blue area).

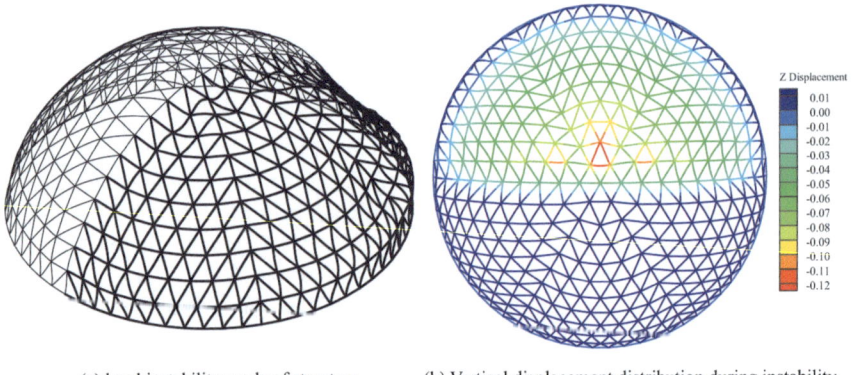

(a) local instability mode of structure     (b) Vertical displacement distribution during instability

**Fig. 2.21** Local instability modes of 50 m span grid shell under half-span uniformly distributed loads (unit: m)

## 2.7.3 Vertex Concentrated Load Mode

The vertical load of 331 N is applied at the vertex (Fig. 2.16) and the load amplitude $\sum F = 331$ N, the total load is equal to the full span uniform load mode. According to Eq. (2.11), the relative change gradient of joint well-formedness can be obtained. The position of each free joint projected to the horizontal plane is taken as the X, Y coordinates, and the *gra_r* of joints is taken as the Z coordinates to draw the spatial distribution map of joint *gra_r* distribution. In order to enhance the expression effect, the *gra_r* of joints is linearly interpolated between discrete joints to generate the spatial distribution cloud map, as shown in Fig. 2.22. The line segment in Fig. 2.22 represents the bar connecting the joints, which can reflect the structural topological connection relationship. According to Fig. 2.22, the *gra_r* of load-bearing joints is significantly lower than that of other joints, indicating that the stiffness degradation of load-bearing joints is the most significant, and its $gra\_r_{\min} = -9.358 \times 10^{-4}$.

The arc length method is adopted to track the elastic stability of the single-layer reticulated shell as shown in Fig. 2.16 under the vertex concentrated load. The stable critical load proportional coefficient $\lambda_{cr} = 1.983 \times 10^3$, and the structural stable bearing capacity $\sum P_{cr} = \lambda_{cr} \times \sum F = 6.56418 \times 10^5$ N are obtained through calculation. The structural instability mode is point instability, as shown in Fig. 2.23.

According to Figs. 2.22 and 2.23, when *gra_r* of individual joints is significantly lower than that of other joints, it indicates that these individual joints have significant softening and the structure shows point instability mode during stable loading; the joint with the most significant softening in Fig. 2.22 has significant instability deformation in Fig. 2.23; the joint in the non-loading area has a high value of *gra_r*, indicating that the softening degree of this area is not as obvious as that of the load-bearing joint and the area remains stable during stable loading, so the area has small deformation in Fig. 2.23b (corresponding to the blue area).

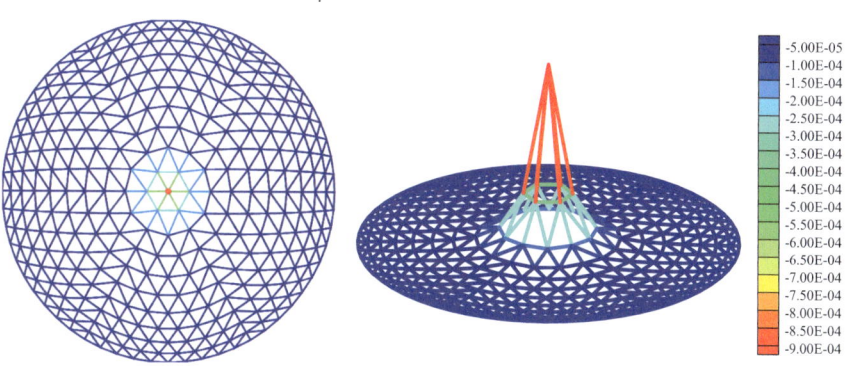

**Fig. 2.22** Nodal *gra_r* distribution of 50 m span single-layer lattice shell structure under vertex-concentrated load

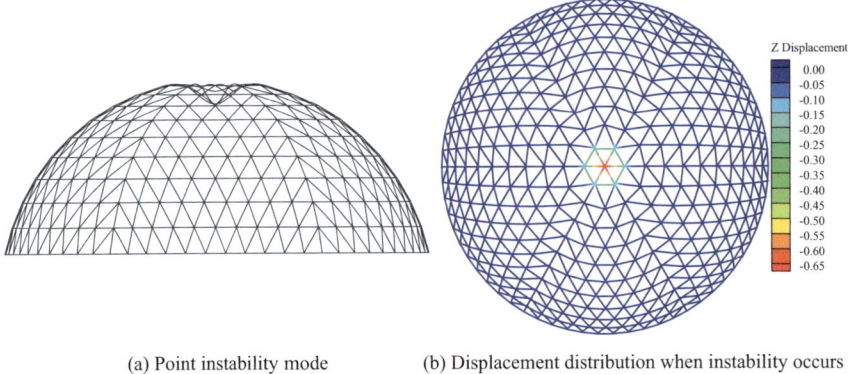

(a) Point instability mode    (b) Displacement distribution when instability occurs

**Fig. 2.23** Point instability mode of 50 m span lattice shell under concentrated load at the vertex (unit: m)

## 2.7.4 Structure Instability Mechanism and Most Unfavorable Load Mode

As an efficient form of space structure, lattice shell is a shape-resistant structure. Through its reasonable curved surface and rod topological relationship, lattice shell makes rod subject to compression under load, and the material utilization is efficient. However, compressive stress also weakens the original stiffness of the structure, making the structure tend to be unstable. The configuration variation characteristics and elastic stability characteristics of the medium span single-layer lattice shell structure shown in Fig. 2.16 under three load modes were studied, and the unstable modes presented include overall unstable, local unstable and point unstable. No matter what kind of unstable mode, the joint $gra\_r$ distribution diagram representing the degree of structural softening is highly consistent with the displacement cloud diagram during structural unstable. In particular, the joint corresponding to $gra\_r_{min}$ is the first to fail in the stable loading process, with significant unstable deformation. This indicates that the extended configuration vulnerability theory can quantitatively measure the degree of weakening of structural stiffness by compressive stress, and the joint with significant softening degree can be accurately identified by calculating the simple $gra\_r$ as an indicator. See Table 2.7 for $gra\_r_{min}$ and $\Sigma P_{cr}$ under the above three load modes. As can be seen from Table 2.7, the lower the $gra\_r_{min}$ is, the lower the structural stable bearing capacity is for the same structure under different load modes. Obviously, the vertex concentrated load is the most unfavorable load mode.

**Table 2.7** The most unfavorable stable load mode of lattice shell structure with 50 m span

| Load mode | $gra\_r_{min}/\times 10^{-6}$ | $\Sigma P_{cr}/\times 10^5$ N |
|---|---|---|
| Vertex concentrated load mode | $-935.8$ | 6.5642 |
| Half span uniform load mode | $-8.731$ | 390.71 |
| Full span uniform load mode | $-4.304$ | 884.83 |

## 2.8 Chapter Summary

In view of the lack of considering load in classical configuration vulnerability theory, this chapter innovatively introduces geometric stiffness matrix, introduces nonlinearity on the basis of first-order analysis, and at the same time, introduces external factors such as load into configuration vulnerability theory, expanding the research scope of the theory. Using the expanded configuration vulnerability theory, the instability mechanism of lattice shell structure is revealed from a new perspective.

The change gradient of joint well-formedness can quantitatively measure the degree of joint softening. Under a given load mode, the joint corresponding to $gra\_r_{min}$ has the most significant degree of softening, and the loading process is the first to fail, which is the key joint to determine the stable bearing capacity of the structure. The joint area with a lower $gra\_r$ has a significant degree of softening, and the loading process is locally unstable. When the $gra\_r$ values of all joints are similar, it indicates that the structure is softened as a whole, and the instability mode is the whole instability mode.

Different load modes, compare the degree of joint softening, and determine the most unfavorable load mode from the stability angle. For spherical shell structure, under the concentrated load mode, the well-formedness of the load point degrades significantly, and the well-formedness distribution is uneven, indicating that part of the structure's load-resistant potential has not been fully developed, and the structure is not reasonably stressed. This load mode is the most unfavorable load mode, and the structure presents a point instability mode. Under the load distribution mode of half span, the joint configuration degradation in the load area is significantly higher than that in the non-load area, and the structure presents a local instability mode of buckling in the load area, and the stable bearing capacity is higher than that in the vertex concentration mode; under the load distribution mode of full span, the $gra\_r$ distribution of each joint is uniform and the peak value is low, and the structure is reasonable, and the stable bearing capacity is the highest.

## References

1. Shen S, Chen X (1999) Stability of reticulated shell structures. Science Press, Beijing
2. Wu X, Blockley DI, Woodman NJ (1993) Vulnerability of structural systems Part 1: rings and clusters. Civ Eng Syst 10(4):301–317

3. Wu X, Blockley DI, Woodman NJ (1993) Vulnerability of structural systems Part 2: failure scenarios. Civ Eng Syst 10(4):319–333
4. Agarwal J, Blockley D, Woodman N (2001) Vulnerability of 3-dimensional trusses. Struct Saf 23(3):203–220
5. Zhu N, Ye J (2013) Structural vulnerability of a single-layer dome based on its form. J Eng Mech 140(1):112–127
6. Ye J, Jiang L, Wang X (2017) Seismic failure mechanism of reinforced cold-formed steel shear wall system based on structural vulnerability analysis. Appl Sci 7(2):182
7. Meek JL, Tan HS (1984) Geometrically nonlinear analysis of space frames by an incremental iterative technique. Comput Methods Appl Mech Eng 47(3):261–282
8. Ragon SA, Gurdal Z, Watson LT (2002) A comparison of three algorithms for tracing nonlinear equilibrium paths of structural systems. Int J Solids Struct 39(3):689–698

**Open Access** This chapter is licensed under the terms of the Creative Commons Attribution-NonCommercial-NoDerivatives 4.0 International License (http://creativecommons.org/licenses/by-nc-nd/4.0/), which permits any noncommercial use, sharing, distribution and reproduction in any medium or format, as long as you give appropriate credit to the original author(s) and the source, provide a link to the Creative Commons license and indicate if you modified the licensed material. You do not have permission under this license to share adapted material derived from this chapter or parts of it.

The images or other third party material in this chapter are included in the chapter's Creative Commons license, unless indicated otherwise in a credit line to the material. If material is not included in the chapter's Creative Commons license and your intended use is not permitted by statutory regulation or exceeds the permitted use, you will need to obtain permission directly from the copyright holder.

# Chapter 3
# Stability Optimization of Single-Layer Lattice Shell Structure Based on Rigid Joints

**Abstract** In the design process of lattice shell structure, the joint is assumed to be ideal rigid, and the bar section design is carried out by considering the three aspects of bar strength, bar stability and overall stiffness. However, for single-layer lattice shell structure, especially for long-span single-layer lattice shell structure, even if the requirements of strength, stiffness and bar stability are met, the overall stability of the structure may not meet the design requirements. However, the current design specification of JGJ7-2010 Technical Specification for Space Grid Structure (Industrial standard of the People's Republic of China, in JGJ7-2010 Technical specification for space grid structure. China Architecture and Building Press, Beijing 2010 [1]) only limits the stability requirements of single-layer lattice shell structure to the checking of stable bearing capacity, and there is no corresponding anti-instability design method. In the study of the size optimization of lattice structure, the overall stability of the structure is generally expressed as a linear eigenvalue problem, and is processed as an optimization constraint condition rather than an optimization target. In the optimization design method that directly takes the structural stability bearing capacity as the optimization target, the optimization object is often a simple articulated structure, or the calculation time is difficult to meet the requirements of practical engineering design. Based on the instability mechanism of lattice shell structure, the optimization model of single-layer lattice shell structure is proposed by calculating the simple $gra\_r_{min}$ to represent the degree of structural stiffness degradation, and the goal is to reduce the instability trend of the structure. The optimization variable is the bar section. In order to consider the actual construction demand, the bar section is taken from the Chinese manufacturing standard GB/T 17395-2008 Seamless Steel Tube Dimensions, Configuration, Weight and Allowable Deviation (National Standard of the People's Republic of China, GB/T 17395-2008 Dimension, shape, weight and allowable deviation of seamless steel pipes. Standards Press of China, Beijing 2008 [2]), which is a discrete variable. The constraint conditions are the design constraint conditions and steel quantity constraint conditions stipulated in the specification. In view of the problems of many variables in the stability optimization model of lattice shell structure and the low efficiency of the traditional optimization algorithm, the random mutation operation in the standard genetic algorithm is improved in this chapter, and the guided genetic algorithm for the stability optimization of large single-layer lattice shell structure is proposed. Finally, the stability optimization design is carried out with three single-layer lattice shell structures of different spans as an example.

## 3.1 Stability Optimization Model of Single-Layer Lattice Shell Structure

Literature [3] pointed out that one of the most important factors in the optimization model is the physical index representing the structural system. A good physical index not only abstracts and condenses the complex structural model into concise mathematical expression, but also takes into account the computational efficiency. The stable critical load Pcr obtained by arc length tracking is currently recognized as the physical quantity representing the structural stability. However, the calculation of lattice shell structure $P_{cr}$ requires nonlinear iteration, which is a huge amount of calculation, and the calculation time increases sharply with the increase of the structure size. In the general optimization process, the calculation of structural response usually accounts for 85–95% of the entire optimization time [4]. If $P_{cr}$ is used to represent the structural stability, the optimization calculation time will not meet the requirements of engineering design.

The instability mechanism of lattice shell structure in Chap. 2 shows that the lower the $gra\_r_{min}$ of the structure is, the more significant the degree of structural stiffness degradation is, and the lower the corresponding stable bearing capacity $P_{cr}$ is. $P_{cr}$ represents the ability of the structure to maintain stability, while $gra\_r_{min}$ represents the structural stability from the opposite side of the structure to maintain stability, namely from the perspective of the structure instability trend. $P_{cr}$ and $gra\_r_{min}$ are the measurements of the structural stability from two different perspectives, both of which are the representatives of the structural stability and have clear physical meanings, but the calculation of $gra\_r_{min}$ is simpler. The stability optimization of large single-layer reticulated shell structure can be realized by using $gra\_r_{min}$ to represent the structural stability.

### 3.1.1 Optimize Goals

The optimization goal is to reduce the instability tendency of the structure and thus increase the stable bearing capacity of the structure. The results in Chap. 2 of this paper show that the larger the $gra\_r_{min}$ of the structure is, the lower the instability tendency of the structure is and the higher the stable bearing capacity $P_{cr}$ is. Therefore, with the maximization of $gra\_r_{min}$ as the optimization goal:

$$\text{Maximize } gra\_r_{min} \qquad (3.1)$$

## 3.1 Stability Optimization Model of Single-Layer Lattice Shell Structure

In the equation, $gra\_r_{min}$ is the minimum value of all non-constrained joints $gra\_r$, as shown in Eq. (3.2).

$$gra\_r_{min} = \min(gra\_r_1, gra\_r_2, \ldots, gra\_r_k, \ldots, gra\_r_n) \qquad (3.2)$$

In the equation, $gra\_r_k$ is the relative change gradient of the well-formedness of joint $k$, as shown in Eq. (2.11); $n$ is the number of unconstrained free joints.

### 3.1.2 Optimize Variables

The single-layer lattice shell structure is composed of bars and joints connecting the bars. In this chapter, the joints are assumed to be ideally rigid, so the bar section is the optimization variable. In order to meet the construction requirements, the candidate section is selected from the Chinese manufacturing standard GB/T 17395-2008 Seamless Steel Tube Dimensions, Shape, Weight and Allowable Deviation [2], which is a discrete variable. The bar section is selected from the candidate section list, and the corresponding outer diameter and wall thickness of the section are determined according to the section number and the candidate section list. Therefore, the optimization variables are as follows:

$$\mathbf{I} = [I_1, I_2, \cdots, I_k, \cdots, I_{nm}] \qquad (3.3)$$

In the equation, $\mathbf{I_k}$ is the sequence number of the $k$-th bar section in the candidate section list; $nm$ is the number of bars; $\mathbf{I}$ is an optimization variable, representing the sequence number of all bar sections in the candidate list, and it is an integer vector.

According to the optimization variable $\mathbf{I}$ and the candidate list, the section size information matrix $\mathbf{S}$ of the bar can be determined:

$$\begin{aligned} \mathbf{S} &= [S_1, S_2, \cdots, S_k, \cdots, S_{nm}]^T \\ \mathbf{S_k} &= [D_k, t_k] \end{aligned} \qquad (3.4)$$

In the equation, $\mathbf{S_k}$ is the section size matrix of the $k$-th bar; $D_k$ and $t_k$ are the outer diameter and wall thickness of the $k$-th bar respectively; $nm$ is the number of bars.

### 3.1.3 Constraint Conditions

The constraint conditions for steel consumption are shown in Eq. (3.5), which gives the upper limit of steel consumption for structural bars. According to the design specifications of GB 50017-2003 Code for Design of Steel Structures [5] and JGJ

7-2010 Technical Specification for Space Grid Structures [1], the design constraint conditions for bars are shown in Eqs. (3.6) and (3.7).

(1) Constraints on the amount of steel used for bars:

$$V_i \leq \eta_V \times V_0 \tag{3.5}$$

In the formula, $V_i$ is the steel amount of the bar part of the structure after the $i$-th optimization step; $V_0$ is the steel amount of the bar part of the initial structure; $\eta_V$ is the adjustment coefficient of the steel amount.

(2) Strength Constraints:

$$\frac{N_i}{A_{ni}} \pm \frac{M_{xi}}{\gamma_x W_{nxi}} \pm \frac{M_{yi}}{\gamma_y W_{nyi}} \leq f \ (i = 1, 2, \ldots, nm) \tag{3.6}$$

In the formula, $N_i$ is the design value of axial force of the $i$-th bar under design load; $A_{ni}$ is the net cross-sectional area of the $i$-th bar; $M_{xi}$ and $M_{yi}$ are the design values of bending moments of the $i$-th bar around the two principal axes under design load respectively; $\gamma_x$ and $\gamma_y$ are the cross-sectional plasticity development coefficients; $W_{nxi}$ and $W_{nyi}$ are the net cross-sectional resistance moments of the $i$-th bar in the two principal axes respectively; $f$ is the design value of the material strength of the bar; $nm$ is the number of bars.

(3) Constraint conditions for stability of compression bars:

$$\frac{N_i}{\varphi_{yi} A_i} + \frac{\beta_{myi} M_{yi}}{\gamma_y W_{yi}(1 - 0.8 N_i / N'_{Ei})} + \eta \frac{\beta_{txi} M_{xi}}{\varphi_{bxi} W_{xi}} \leq f \ (i = 1, 2, \ldots, ncm) \tag{3.7}$$

In the equation, $A_i$ is the gross section area of the $i$-th bar; $W_{xi}$ and $W_{yi}$ are the gross section resistance moments of the $i$-th bar in the two principal axis directions respectively; $\varphi_{yi}$ is the axial compression stability coefficient of the $i$-th bar; $\eta$ is the section influence coefficient; $\beta_{myi}$ and $\beta_{txi}$ are the equivalent bending moment coefficients of the $i$-th bar in the two principal axis directions respectively; $\varphi_{bxi}$ is the overall stability coefficient of the $i$-th bar under uniform bending; $N'\ Ei$ is the Euler critical force of the $i$-th bar; $ncm$ is the number of compression bars.

## 3.2 Stability Optimization Algorithm for Single-Layer Lattice Shell Structure

### 3.2.1 Canonical GA

Canonical GA is a heuristic probabilistic search algorithm which simulates the genetic and evolutionary process of population in natural environment according to Darwin's theory of evolution. This method has been widely used in the optimization of rod structure. The main steps of the implementation of the canonical genetic algorithm are as follows:

(1) Encode the optimized object and generate the initial population;
(2) Calculate the fitness of each individual in the population;
(3) Based on the individual fitness, select the individual with high fitness and eliminate the individual with low fitness through selection operation, simulating the mechanism of "natural selection and survival of the fittest" in nature;
(4) Cross and randomly mutate the selected population to simulate the reproduction and gene mutation of the population in nature;
(5) Truncated judgment, if the optimization meets the stop condition, then stop evolution, otherwise go to step (2).

Through crossover operation, the genes in the parent individual are inherited to the offspring; through random mutation operation, new genes are generated, making the offspring individuals move randomly in the design space to achieve global search; finally, through selection operation, the excellent individual with high fitness value is retained, and the inferior individual is eliminated, making the result tend to the global optimal.

In the case of space lattice shell structure, the large number of rod components leads to a large number of optimization variables, a long number of chromosome genes, and many design constraints, resulting in a large amount of calculation. Therefore, in order to obtain ideal optimization results, the standard genetic algorithm requires a considerable number of populations, reasonable mutation probability, crossover probability and sufficient evolutionary generations. But this often requires huge calculation time and computing space. At the same time, for the optimization of different lattice shell structures under different load conditions, the selection of reasonable parameters can only rely on experience. Therefore, the standard genetic algorithm has significant parameter sensitivity and problem dependence. Secondly, when dealing with the optimization problem of rod structure, the standard genetic algorithm cannot guarantee the continuity of parent and offspring individuals. Random mutation of alleles can only ensure that the chromosome is continuous, but cannot ensure that the corresponding phenotype of the chromosome is continuous [3]. At the same time, this completely random mutation may also destroy the original excellent gene segment, which makes it difficult for the population to evolve to the optimal solution when the optimization problem is complex. Therefore, in order to solve the stable optimization problem, it is necessary to develop an appropriate optimization strategy so that the genetic algorithm can steadily find the optimal solution in the design space.

### 3.2.2 Guided GA

Different from the random mutation in the standard genetic algorithm, the guided GA proposed in this section makes full use of the results of joint well-formedness analysis, combines with the instability mechanism of lattice shell structure, identifies the key bars that determine the stability of lattice shell structure, and carries out directional mutation on the key bars. The directional mutation improves the evolution efficiency of the population and ensures the continuity of the population evolution. The specific steps of the guided genetic algorithm are as follows:

(1) **Coding and population initialization**: Binary coding scheme suitable for discrete variable problem is adopted in this paper. The $ns$ cross-sections within the range of values of the optimization variable are numbered in ascending order according to their areas, and the number of the cross-section is denoted as $S$ ($S = 1, 2 \ldots ns$). For the cross-section numbered as $S$, its area is $A^S$ and its cyclotron radius is $i^S$. According to the binary coding rules, the decimal number of the bar is converted into binary coding to generate chromosomes. The initial population satisfying the steel amount is randomly generated.

(2) **Joint well-formedness analysis and individual fitness calculation**: decode a chromosome encoded in binary code to obtain its corresponding phenotype, namely the corresponding lattice shell structure. Under given constraints and stable load conditions, carry out well-formedness analysis of the lattice shell structure according to the method in Chap. 2, and obtain $gra\_r$ of each joint and $gra\_r_{min}$ of the structure. Since the optimization goal is to maximize the $gra\_r_{min}$ of the structure, the configuration fitness function is as follows:

$$f(x) = \frac{1}{|gra\_r_{min}|} \qquad (3.8)$$

In the formula, $x$ is a chromosome in the current population; $f(x)$ is the fitness of the individual without considering the constraints. For the design constraints in the optimization model, this paper adopts the penalty function method to transform the constrained optimization problem into unconstrained optimization problem. This paper adopts the penalty function proposed by Gen and Cheng [6], which is constructed as follows:

$$p(x) = 1 - \frac{1}{m} \sum_{i=1}^{m} \left[ \frac{\Delta b_i(x)}{\Delta b_i^{max}} \right]$$

$$\Delta b_i(x) = max\{ 0, g_i(x) - b_i(x) \}$$

## 3.2 Stability Optimization Algorithm for Single-Layer Lattice Shell Structure

$$\Delta b_i^{max} = \max\{\varepsilon, \Delta bi(x)\} \tag{3.9}$$

In the formula, $p(x)$ is the penalty function value of individual $x$, with the value range between 0 and 1; $\Delta b_i(x)$ is the violation amount of individual $x$ to the $i$ ($i = 1, 2$) constraint condition; $\Delta b_i^{max}(x)$ is the maximum violation amount of individuals in the current population to the $i$ constraint condition; $m$ is the number of constraints violated by individual $x$; $g_i(x)$ is the calculated value of individual $x$ to the $i$ constraint condition; $\varepsilon$ is a small positive number that avoids division by 0. When an individual does not violate the constraint condition, $p(x) = 1$; when an individual violates the constraint condition, $0 < p(x) < 1$; the more significant the violation of the constraint condition is, the more $p(x)$ approaches 0.

After considering the design constraint condition, the final individual's fitness function is as follows:

$$F(x) = p(x)f(x) \tag{3.10}$$

(3) **Selecting excellent individuals**: Ranking all individuals in the population according to their fitness. The best 1/4 individuals are copied twice, and the middle 1/2 individuals are copied once. This truncation selection is a common selection method in genetic algorithms, which can ensure that the best individual genes are retained.

(4) **Crossover operation**: select several paired parental chromosomes from the population according to a certain crossover rate, exchange part of genes with each other according to the single-point crossover method, so as to form two new offspring chromosomes. Crossover operation simulates the characteristics of biological genetics, so that good genes can be inherited to the next generation of population. Crossover operation can promote the search ability of the solution space, avoid local optimum, and improve the convergence speed.

(5) **Identification of key members**: Strengthen the section of weak members to improve their stable bearing capacity; meanwhile, reduce the section of rigid redundant members to make the distribution of structural rigidity reasonable and satisfy the constraint conditions of steel consumption. Identification of weak members and redundant members is as follows:

(5.1) **Identification of weak members**: Select $nvj$ joints with the lowest $gra\_r$. In stable tracking, these joints are generally the first to fail, and are defined as the set of weak joints $\{J_v\}$ Among all the members connected with weak joint $J_{v,i}$ ($i = 1, 2, ..., nvj$), the member with the smallest cross-sectional area is weak member $b_{v,i}$, the section of which is numbered $S_{v,i}$, the cross-sectional area is $A_{v,i}$, and the corresponding cyclotron radius is $i_{v,i}$.

(5.2) **Identification of redundant members**: Select $nrj$ joints with the highest $gra\_r$. In stable tracking, these joints are generally stable, and are defined as the set of redundant joints $\{J_r\}$. Among all the members connected with redundant joint $J_{r,i}$ ($i = 1, 2, ..., nrj$), the member with the largest cross-sectional area is redundant member $b_{r,i}$, the section of which is numbered $S_{r,i}$, and the cross-sectional area is $A_{r,i}$.

(6) **Directed mutation of key members:** adjust weak members and redundant members targeted, so that the population gradually approaches the optimal solution. The specific methods are as follows:

(6.1) **Weak variants**: For weak member $b_{v,i}$, its section is numbered $S_{v,i}$ and its section area is $A_{v,i}$. Since the list of candidate sections is arranged in ascending order by area, the sections after $S_{v,i}$ have areas larger than $A_{v,i}$, but their gyroscopic radii are not necessarily larger than $i_{v,i}$. The first section after $S_{v,i}$ with a gyroscopic radius not less than $i_{v,i}$ is selected and assigned to member $b_{v,i}$ The new section number $S_{v,i}^{new}$ is shown in Eq. (3.11).

$$S_{v,i}^{new} = S_{v,i} + R \qquad (3.11)$$

In the equation, $R$ is an integer not less than 1, which should satisfy the requirements as follows:

$$\begin{aligned} i^{S_{v,i}+k} &< i_{v,i} \, (k = 1, 2, \ldots, R-1) \\ i^{S_{v,i}+R} &\geq i_{v,i} \end{aligned} \qquad (3.12)$$

In the equation, $i_{v,i}$ is the sectional gyroscopic radius of rod $b_{v,i}$ in the current optimization step.

(6.2) **Mutation of redundant rod**: for redundant rod $b_{r,i}$, its section is numbered $S_{r,i}$.

Since the list of candidate sections is arranged in ascending order by area, the section before $S_{r,i}$ is slightly smaller than this section. See Eq. (3.13) for the new section number $Snew\ r,i$.

$$S_{r,i}^{new} = S_{r,i} - 1 \qquad (3.13)$$

(7) **Threshold judgment**: if the evolution has reached enough generations, stop optimization; otherwise, go to step

The evolutionary operation of the guided genetic algorithm is almost the same as that of the standard genetic algorithm, which are both calculating individual fitness, selection operation, crossover operation and mutation. The standard genetic algorithm adopts random mutation, which takes the existing point in the design space (corresponding to an individual in the population) as the benchmark and conducts a random search in the surrounding design space according to one direction (corresponding to the mutated individual).If the search results approach the optimal solution, they will be retained in the selection operation; if the search results are far from the optimal solution, they will be eliminated in the selection operation. The guided genetic algorithm adopts directional mutation, which makes the mutation operation gradually approach the optimal solution by strengthening the weak members and weakening the redundant members, thus greatly improving the evolution rate.

The process of the guided genetic algorithm is shown in Fig. 3.1.With the beginning of optimization, the structure will gradually tend to optimization, which is manifested in the great reduction of the number of weak joints and the maximum mining of the potential of the overall load resistance of the structure. At this time, if the value of the weak joint number *nvj* is large, the optimization program will process some non-weak joints as weak joints, which is not conducive to the final optimization results. Therefore, the number of weak joints can be selected smaller, generally recommended not to be greater than 10.

## 3.3 Optimization Example 1: 22 m Span Single-Layer Reticulated Shell

### *3.3.1 Basic Information of Structure*

As shown in Fig. 3.2, the small-span single-layer K6 lattice shell is composed of 6 identical sectors, each of which is shown in the black bold part in Fig. 3.2. The K6 lattice shell spans 22 m, has a beam height of 11 m, and is surrounded by fixed supports. The structure has a total of 462 bars and 169 joints. The material is Q345, and the steel amount for bars is 1.41 $m^3$, equivalent to 29.31 kg/$m^2$. According to the Technical Specification for Space Grid Structures JGJ7-2010 [1], the full-span uniform load mode is the stable control load mode of single-layer spherical lattice shell structure.

The optimization variable is the section of the bar, and the value range is the section with the outer diameter of 89–406 mm and the wall thickness not greater than 20 mm in GB/T 17395-2008 Seamless Steel Tube Dimensions, Configuration, Weight and Allowable Deviation [2], a total of 196 candidate sections. The candidate sections almost include all suitable sections. In Eq. (3.5), $\eta_v = 1.0$, that is, the upper limit of the steel amount of the structural bar is kept at 1.41 $m^3$. For the initial structure, on the premise of ensuring that the overall steel amount is close to the limit value of the steel amount, a section is randomly selected from the list of candidate sections and assigned to each bar. The steel amount of the initial structure is 1.4060 $m^3$.

### *3.3.2 Stable Optimization by Standard Genetic Algorithm*

The standard genetic algorithm is adopted to carry out stable optimization of K6 reticulated shell shown in Fig. 3.2 under full span uniform load. The population sizes are 5 and 100, respectively. According to the conventional value ranges of

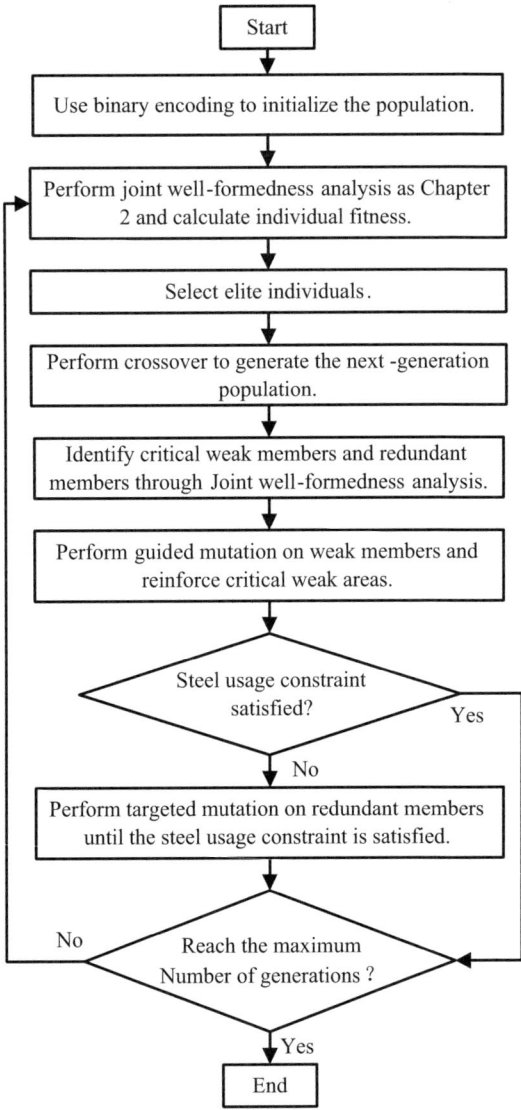

**Fig. 3.1** Guided genetic algorithm process

crossover rate and mutation rate in the standard genetic algorithm [7] and combined with calculation experience, the crossover rate is set as 0.2 and the mutation rate as 0.02. The standard genetic algorithm runs on a computer with WIN7 operating system, which is configured with Inter(R) Core(TM) i7-4790 K CPU@ 4.00 GHz and 32 GB memory. In the case of population size of 5 and 100, the evolution process of structure $P_{cr}$ and $gra\_r_{min}$ in the standard genetic algorithm is shown in Figs. 3.3 and 3.4, respectively.

## 3.3 Optimization Example 1: 22 m Span Single-Layer Reticulated Shell

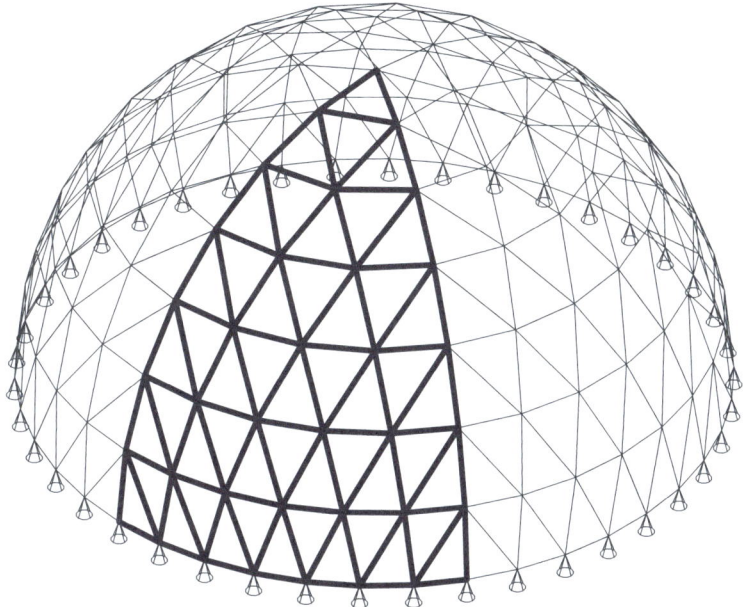

**Fig. 3.2** 22 m span single layer spherical reticulated shell

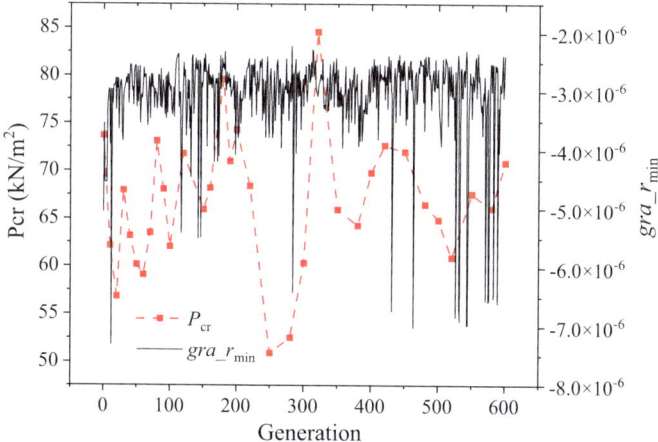

**Fig. 3.3** Evolutionary history with population size of 5

Figures 3.3 and 3.4 show that when the population size is small, the optimization algorithm has no obvious effect on improving $gra\_r_{min}$, and $gra\_r_{min}$ fluctuates widely throughout the optimization process. When the population size is enlarged to 100, the standard genetic algorithm can effectively improve $gra\_r_{min}$, and $gra\_r_{min}$ fluctuates narrowly throughout the optimization process, and tends to converge

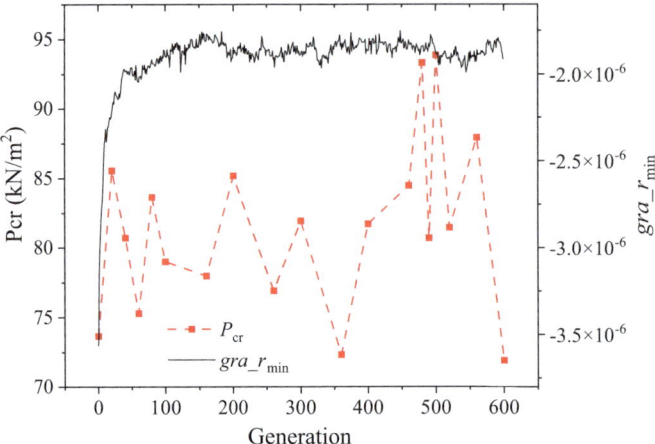

**Fig. 3.4** Evolutionary history with population size of 100

in the late optimization process. Accordingly, when the population size is 100, the optimized structure has a $P_{cr} = 93.31$ kN/m², which is higher than the optimized structure with a population size of 5 and a $P_{cr} = 84.55$ kN/m². When the population size is 5, the maximum fluctuation range of 33.65 kN/m²; when the population size is 100, the maximum fluctuation range of $P_{cr}$ is 21.41 kN/m². However, although the fluctuation range of $P_{cr}$ is slightly reduced with the enlargement of the population size, the optimization effect on the structural stability bearing capacity is not significant, because: (1) the optimization variable is the discrete bar section, and random mutation cannot guarantee that the chromosomes before and after mutation are continuous in phenotype (i.e. the lattice shell structure corresponding to the chromosome decoding) [3]; (2) the lattice shell stability is very sensitive to the changes in the bar section, especially some key bar section. Take the optimization results of the 500th generation with a population size of 100 as an example: when the structure is unstable, its instability mode is shown in Fig. 3.5. In Fig. 3.5, the instability deformation of the structure is concentrated on the 6 main rib joints, presenting a point instability mode. The specifications of the instability joints and their connected bars are shown in Fig. 3.6. Figure 3.6 shows that the main rib bar on the top (outer diameter: 89 mm, wall thickness: 1.6 mm) is significantly weaker than other bars, leading to the joint instability in the process of stable loading. The reason for this situation is the random mutation operation in the standard genetic algorithm. For the bar to be mutated, the random mutation operation is equivalent to randomly selecting a different section from the list of candidate sections and assigning it to the bar. This random mutation operation can neither guarantee the requirement of *continuous change* of the section of the bar before and after mutation, nor ignore the requirement of *continuous* structural stiffness in structural design, resulting in a huge difference in the section size of the bars intersecting at a point. If the specification of the main rib bar on the top is set to be the same as that of the main rib bar on

## 3.3 Optimization Example 1: 22 m Span Single-Layer Reticulated Shell

the bottom, the structural stable bearing capacity will be increased by 8.5% on the premise of almost no significant increase in the amount of steel used in the structure. Therefore, it is precisely because of the "discontinuity" caused by random mutation and the sensitivity of shell stability to bars that the standard genetic algorithm cannot effectively improve the structural stable bearing capacity even on the premise of maximizing $gra\_r_{min}$.

Expanding the population size can alleviate the negative impact of the above problems in theory. However, for the stability optimization problem of the single-layer lattice shell structure, each optimization variable has a wide range of values and the number of optimization variables is large, resulting in a high-dimensional problem of stability optimization, and the reasonable population size is necessarily very large. The time required by large-scale population is difficult to meet the actual demand of the project. For the small span lattice shell structure, when the population size increases from 5 to 100, the optimization time increases from 44 to 780 min. With the expansion of the population size, the calculation time will increase geometrically.

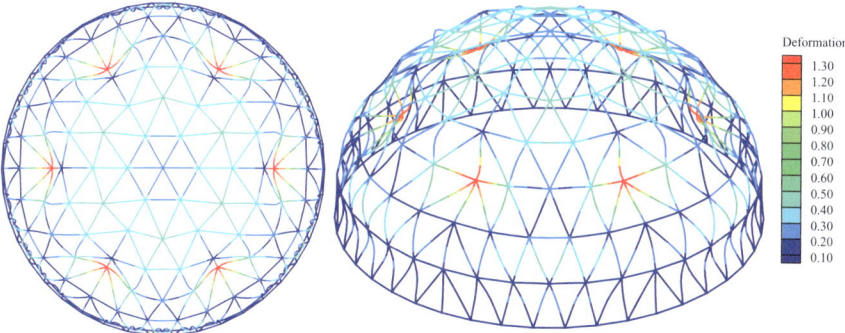

**Fig. 3.5** Structure optimized by standard genetic algorithm and its instability mode (deformation enlarged by 2 times, unit: m)

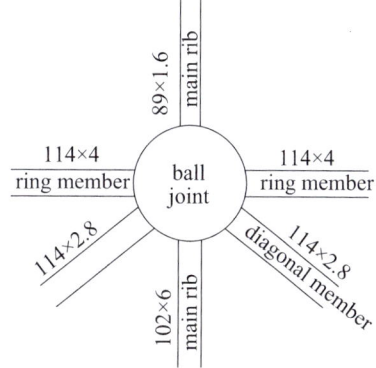

**Fig. 3.6** Instability joint and its connected members

### 3.3.3 Guided Genetic Algorithm for Solving Stable Optimization

The guided genetic algorithm proposed in this paper is used to carry out stable optimization of K6 single-layer lattice shell shown in Fig. 3.2 under full span uniform load. In the optimization algorithm, the number of weak joints is $nvj = 7$, the crossover rate is still 0.2, and the population number is 5. The four cases of the number of redundant joints $nrj$ is 37, 61, 91 and 127 are optimized, and the $gra\_r_{\min}$ optimization process of the corresponding structure is obtained. For the lattice shell structures of previous generations in the optimization process, the arc length method is used to track the evolution process of the stable bearing capacity $P_{cr}$ of the perfect structure. At the same time, the defects are considered according to the Technical Specification for Space Grid Structures JGJ7-2010 [1], and the evolution process of the stable bearing capacity $P_{cr}^{im}$ of the defective structure is obtained. In the four cases, the evolution process of $gra\_r_{\min}$, $P_{cr}$ and $P_{cr}^{im}$ of the corresponding structure are shown in Figs. 3.7, 3.8, 3.9 and 3.10 respectively.

Different from the standard genetic algorithm, the guided genetic algorithm can evolve to the optimal solution quickly in a small population scale. After evolved to the optimal solution, even if the evolution continues, it can also converge to the optimal solution. At the same time, the evolution generations required by the guided genetic algorithm are significantly less than the standard genetic algorithm. Taking $nrj = 91$ as an example to explain the optimization process of the guided genetic algorithm in detail: when $nrj = 91$, the optimization process of $P_{cr}$, $P_{cr}^{im}$ and $gra\_r_{\min}$ is shown in Fig. 3.9. Figure 3.9 shows that in the first 20 steps, $gra\_r_{\min}$ is rapidly increased and basically reaches the optimal solution, and the stable bearing capacity of the structure is also greatly improved at this stage; thereafter, due to the constraint of

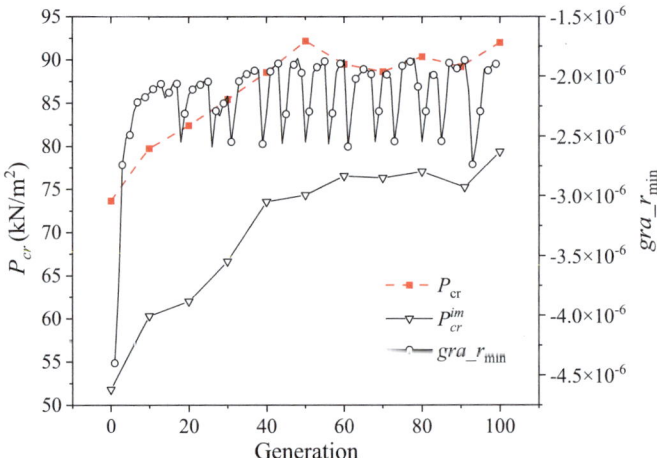

**Fig. 3.7** Optimization process when $nrj = 37$

## 3.3 Optimization Example 1: 22 m Span Single-Layer Reticulated Shell

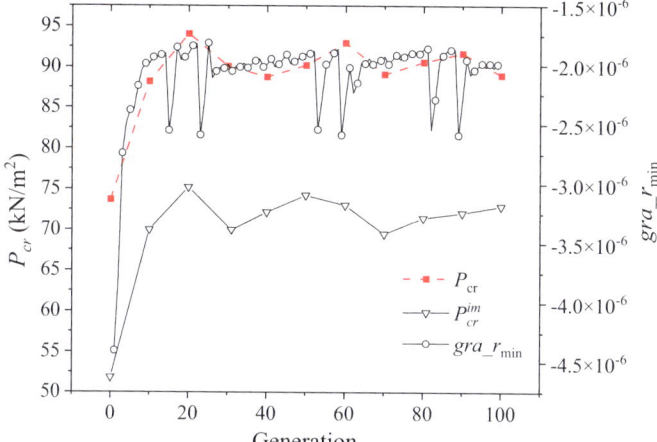

**Fig. 3.8** Optimization process when $nrj = 61$

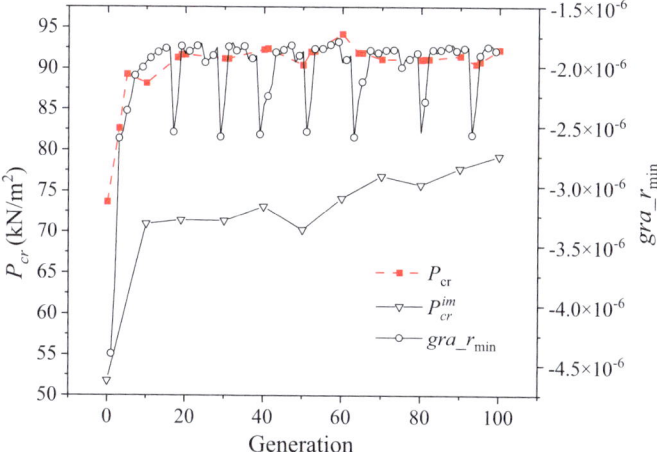

**Fig. 3.9** Optimization process when $nrj = 91$

steel quantity, $gra\_r_{min}$ fluctuates in the process of adjusting the stiffness redundant bar. The stable tracking of the optimization results of $gra\_r_{min}$ fluctuation shows that the stable bearing capacity is not reduced as a result. In the whole subsequent optimization process, the stable bearing capacity of the structure basically remains unchanged. With the bar stiffness distribution tending to be reasonable, even after considering the influence of defects on the stable bearing capacity according to the specification, $P_{cr}^{im}$ is also gradually improved. Because the structure is symmetric in rotation, the load distribution is also symmetric, so the distribution of the bar in the 6 sectors of the optimized structure is completely the same. Taking the black bold

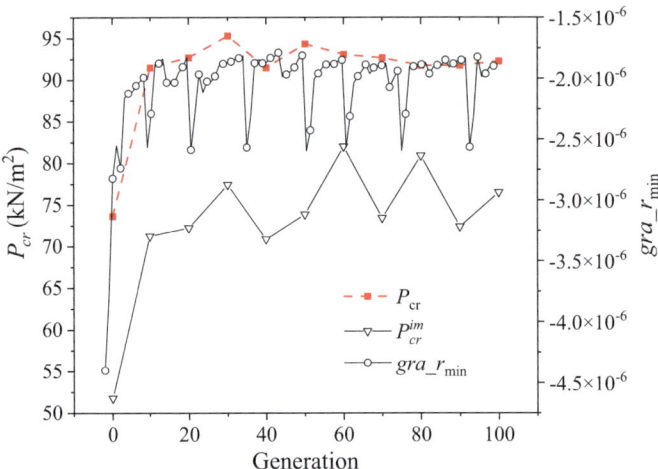

**Fig. 3.10** Optimization process when $nrj = 127$

sector in Fig. 3.2 for illustration, the distribution of the bar in the optimized structure is shown in Fig. 3.11. There are 7 types of sections in the optimized structure, and the steel amount used for the bars is 1.40 m³, meeting the constraint condition of steel amount used for the bars. The stable bearing capacity of the optimized structure is $P_{cr} = 94.22$ kN/m², which is 27.9% higher than that of the initial structure $P_{cr}$. After considering the defects, the stable bearing capacity of the optimized structure is $P_{cr}^{im} = 79.29$ kN/m², which is 52.8% higher than that of the initial structure $P_{cr}^{im}$. Keeping other conditions unchanged and changing the number of redundant joints, the guided genetic algorithm can evolve to a similar optimization solution and remain stable (see Figs. 3.7, 3.8 and 3.10), with only a slight difference in the optimization time.

### 3.3.4 Comparison of Optimization Algorithms

The calculation results of the standard genetic algorithm and the guided genetic algorithm are shown in Table 3.1. The optimization algorithms are run on the same computer with WIN7 operating system, which is configured with Inter (R) Core (TM) i7-4790 K CPU @ 4.00 GHz and 32 GB memory. It can be concluded from Table 3.1 that: (1) compared with the standard genetic algorithm, the guided genetic algorithm requires a small population size and fewer evolution generations; (2) no matter what the value of the parameter $nrj$ in the guided genetic algorithm is, the optimized lattice shell structure has good stable bearing capacity, and the stable bearing capacity of the optimized lattice shell is basically the same, which indicates that the guided genetic algorithm does not depend on the optimization parameter and has good robustness;

## 3.3 Optimization Example 1: 22 m Span Single-Layer Reticulated Shell

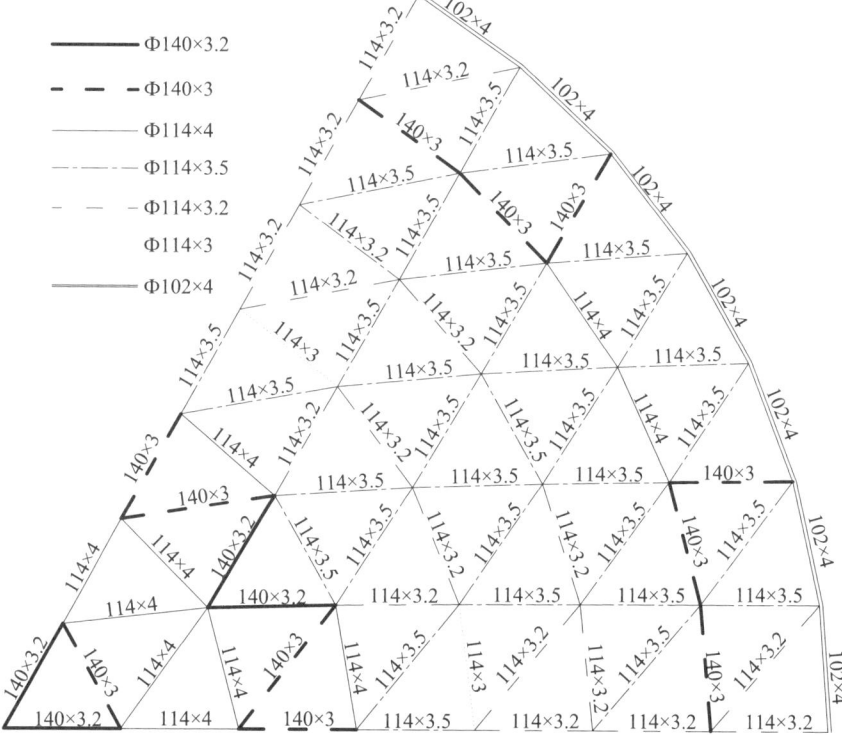

**Fig. 3.11** Sections of 22 m span lattice shell structure after optimization

(3) for the lattice shell structure, the time required by the guided genetic algorithm is about 10 min, which is significantly less than the standard genetic algorithm.

The evolution of structural $P_{cr}$ under different optimization parameters of the two algorithms is shown in Fig. 3.12. The random mutation mechanism in the standard genetic algorithm cannot guarantee the continuity of individuals before and after

**Table 3.1** Comparison of optimization results of 22 m span gridshell

| Parameters for calculation | GA | GA | GGA | | | |
|---|---|---|---|---|---|---|
| | | | $nrj = 37$ | $nrj = 61$ | $nrj = 91$ | $nrj = 127$ |
| Population size | 5 | 100 | 5 | 5 | 5 | 5 |
| Evolutionary algebra | 600 | 600 | 100 | 100 | 100 | 100 |
| $P_{cr}(kN/m^2)$ | 80.73 | 93.84 | 92.15 | 92.96 | 94.22 | 94.36 |
| Amount of steel used ($m^3$) | 1.39 | 1.40 | 1.40 | 1.40 | 1.40 | 1.39 |
| $gra\_r_{min}(10^{-6})$ | −2.584 | −1.846 | −1.850 | 1.814 | −1.790 | −1.789 |
| Time taken (min) | 44 | 780 | 5 | 7 | 10 | 11 |
| Time taken (min/generation) | 0.0733 | 1.3 | 0.05 | 0.07 | 0.1 | 0.11 |

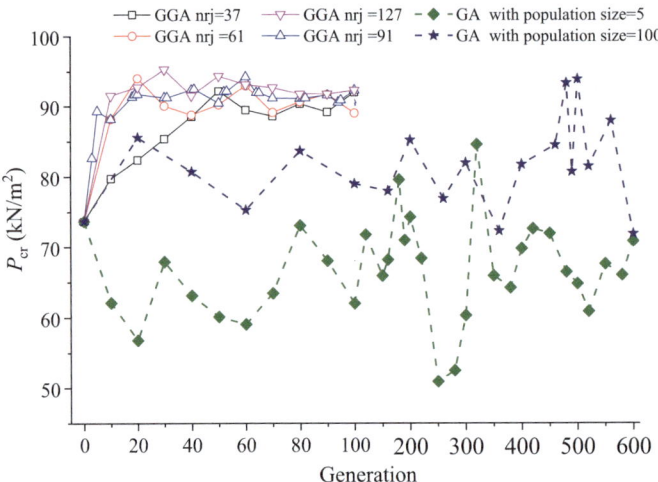

**Fig. 3.12** Optimization history of each optimization algorithm

mutation, so the optimization process of structural $P_{cr}$ is unstable. Even if the standard genetic algorithm can search for the optimal solution, the random mutation will destroy the excellent gene, causing the oscillation of $P_{cr}$. Unlike the standard genetic algorithm, the guided genetic algorithm can ensure that the final optimization results converge to the optimal solution. In the guided genetic algorithm, the difference of the optimization parameter nrj will lead to the difference of the evolution rate of $P_{cr}$. However, no matter what the value of *nrj* is, $P_{cr}$ basically remains stable after 60 optimization steps; at the same time, no matter what the value of nrj is, the value of $P_{cr}$ of the optimization results is basically the same.

## 3.4 Optimization Example 2: 50 m Span Single-Layer Reticulated Shell

### 3.4.1 Basic Information of Structure

As shown in Fig. 3.13, the medium span K6 single-layer lattice shell has a span of 50 m, a height of 20 m, and fixed supports around. The structure has a total of 1122 bars and 397 joints. The lattice shell structure is composed of 6 identical sectors, and a sector is shown as the bold bars in Fig. 3.13.

The steel used for the structure is Q345. The standard value of dead load is 2.55 kN/m², the standard value of live load is 0.50 kN/m², and the design value of load is 3.93 kN/m². According to the GB 50017-2003 Code for Design of Steel Structures [5] and the JGJ7-2010 Technical Regulations for Space Grid Structures

## 3.4 Optimization Example 2: 50 m Span Single-Layer Reticulated Shell

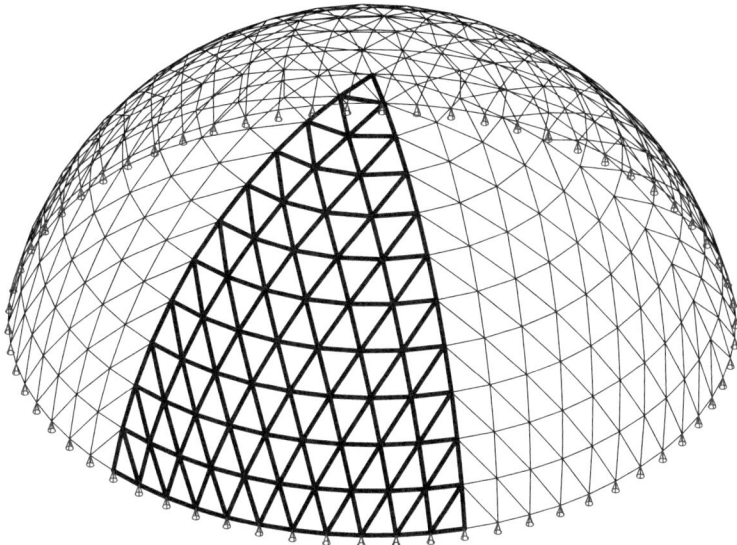

**Fig. 3.13** 50 m span single layer spherical reticulated shell

[1], after the full stress design, the steel amount of the initial structure bar is 6.08 m³, equivalent to 24.46 kg/m², and the initial structure meets the requirements of stress and component stability. The specific mechanical indicators are shown in Table 3.2 According to the JGJ7-2010 Technical Regulations for Space Grid Structures [1], the full span uniform load mode is the stable control load mode of the single layer spherical lattice structure. In the full span uniform load mode, the initial structure $gra\_r_{min} = -6.43358 \times 10^{-6}$; the critical load $P_{cr}$ of the improved structure is $P_{cr} = 26.66$ kN/m² after the elastic stability tracking of the arc length method; the critical load $P_{cr}^{im}$ of the defective structure is $P_{cr}^{im} = 12.59$ kN/m², after the defects are considered according to the JGJ7-2010 Technical Regulations for Space Grid Structures [1]. For the initial structure, according to the safety factor method in the JGJ7-2010 Technical Regulations for Space Grid Structures [1] (the safety factor is 4.2), $P_{cr}^{im}/4.2 = 2.99$ kN/m² $< 2.55 + 0.50 = 3.05$ kN/m², the stability checking is not passed, and the stability checking is not passed. The above examples show that the lattice structure obtained by the traditional method may not pass the stability checking.

**Table 3.3.2** Mechanical indexes of initial structure of 50 m span reticulated shell

| Mechanical index | Computed value | Limit value |
| --- | --- | --- |
| Maximum stress (N/mm²) | 112.06 | 310 |
| Maximum plane stable stress (N/mm²) | 115.63 | 310 |
| Maximum plane stability stress (N/mm²) | 212.88 | 310 |

The optimization variable is the section of the bar, and the value range is the section with the outer diameter of 89–406 mm and the wall thickness not greater than 20 mm in GB/T 17395-2008 Seamless Steel Tube Dimensions, Configuration, Weight and Allowable Deviation [2], a total of 196 candidate sections. The candidate sections almost include all suitable sections. In Eq. (3.5), $\eta_v = 1.0$, that is, the upper limit of the steel amount used for the structural bar is 6.08 m$^3$.

### 3.4.2 Stable Optimization by Standard Genetic Algorithm

Standard genetic algorithm is adopted to carry out stable optimization of medium span K6 single-layer reticulated shell as shown in Fig. 3.13 under full span uniform load. The population sizes are 5 and 100, respectively. According to the conventional value ranges of crossover rate and mutation rate in standard genetic algorithm [7] and combined with calculation experience, the crossover rate is set as 0.2 and the mutation rate as 0.002. The standard genetic algorithm runs on a computer with WIN7 operating system, which is configured with Inter(R) Core(TM) i7-4790 K CPU@ 4.00 GHz and 32 GB memory. In the case of population size of 5 and 100, the evolution process of structure $P_{cr}$ and $gra\_r_{min}$ in the standard genetic algorithm is shown in Figs. 3.14 and 3.15, respectively.

Figure 3.14 shows that for the single-layer lattice structure with medium span, the standard genetic algorithm can improve $gra\_r_{min}$ when the population size is small, but the optimization effect is not as good as that when the population size is large (Fig. 3.15). Meanwhile, $gra\_r_{min}$ shows a very significant fluctuation in the optimization process. When the population size is 5, the Pcr of the structure is slightly increased with the increase of $gra\_r_{min}$ in the first 30 optimization steps,

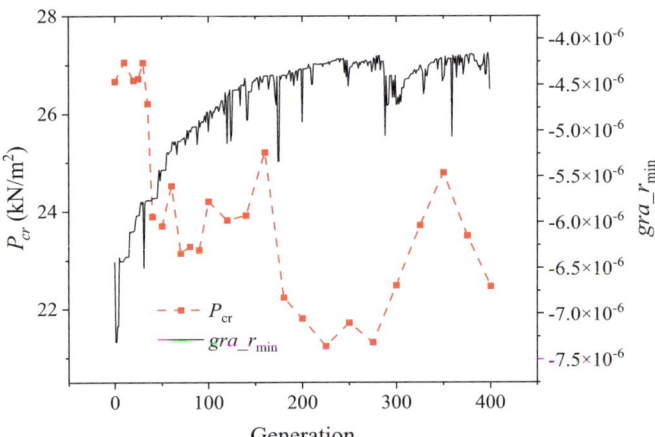

**Fig. 3.14** Evolutionary history of structure with population size of 5

### 3.4 Optimization Example 2: 50 m Span Single-Layer Reticulated Shell

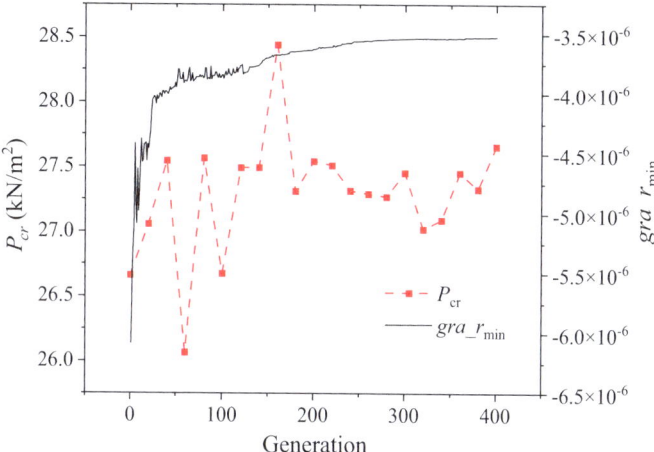

**Fig. 3.15** Evolutionary history of structure with population size of 100

and the $P_{cr}$ of the optimized structure is 27.05 kN/m², slightly larger than that of the initial structure. However, with the optimization progress, the destructive effect of random mutation on the excellent chromosome gradually becomes prominent, and the population size is not enough to alleviate the negative impact. Therefore, the Pcr generally shows a downward trend in the later optimization process, and the stable bearing capacity of the final structure is lower than that of the initial structure. When the population size is expanded to 100, Fig. 3.15 shows that the standard genetic algorithm can effectively improve $gra\_r_{min}$, and $gra\_r_{min}$ fluctuates slightly in the whole optimization process, and tends to converge in the later optimization process. However, the $P_{cr}$ of the optimized structure is 28.44 kN/m², slightly larger than that of the initial structure ($P_{cr}$ = 26.66 kN/m²). In the later optimization process, the stable bearing capacity of the structure is only slightly larger than that of the initial structure. Combining Figs. 3.14 and 3.15, for the single-layer lattice structure, the stable bearing capacity of the structure is only increased from 27.05 to 28.44 kN/m² with the increase of population size, and the optimization effect is not obvious, but the calculation time increases from 104 to 3166 min (nearly 53 h). Therefore, for the medium span single-layer reticulated shell structure, increasing the population size cannot alleviate the discontinuous problem caused by random mutation, and a new optimization algorithm is urgently needed to solve it.

### 3.4.3 Guided Genetic Algorithm for Solving Stable Optimization

The guided genetic algorithm proposed in this paper was used to carry out stable optimization of the medium span K6 single-layer reticulated shell as shown in Fig. 3.13

under full span uniform load. In the optimization algorithm, the number of weak joints $nvj = 7$, the crossover rate is still 0.2, and the population number is 5. The three cases of redundant joints nrj being 91, 127 and 169 were optimized, and the evolution process of the corresponding structure $P_{cr}$ and $gra\_r_{min}$ was shown in Figs. 3.16, 3.17, and 3.18.

Compared with the small-span single-layer lattice shell structure in Sect. 3.3, the optimization object in this section is the medium-span single-layer lattice shell structure, which has an order of magnitude more bar than the lattice shell structure in Sect. 3.3. Nevertheless, the guided genetic algorithm can also evolve to the optimal

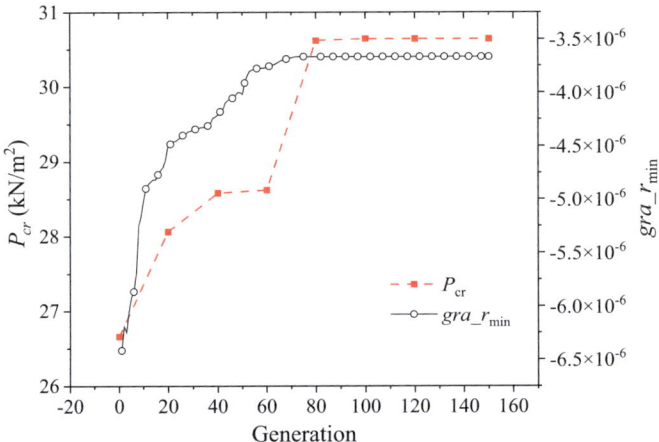

**Fig. 3.16** Optimization process of structure when $nrj = 91$

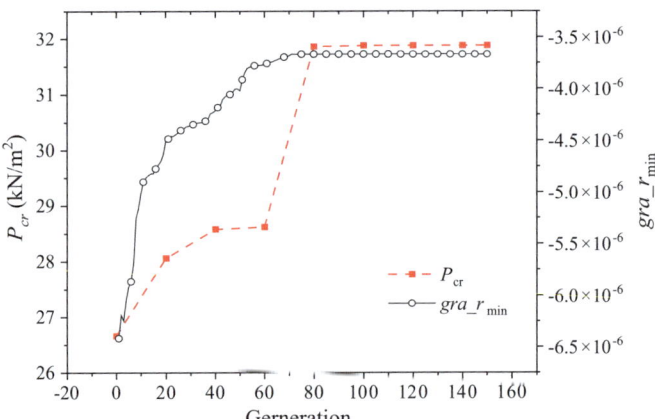

**Fig. 3.17** Optimization process of structure when $nrj = 127$

## 3.4 Optimization Example 2: 50 m Span Single-Layer Reticulated Shell

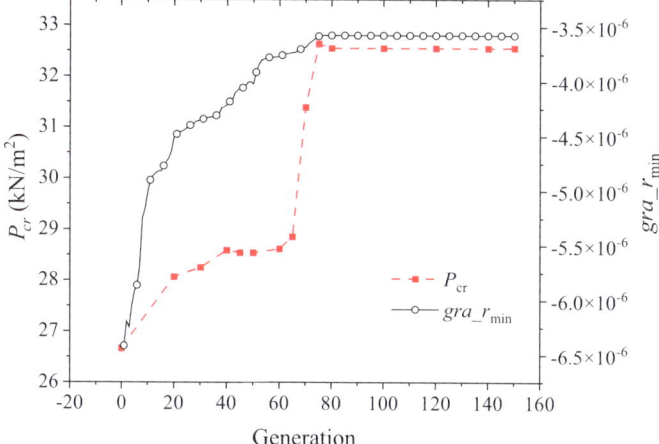

**Fig. 3.18** Optimization process of structure when $nrj = 169$

solution quickly and converge to the optimal solution without increasing the population size (i.e. the population size is still 5). Meanwhile, the evolution generations required by the guided genetic algorithm are obviously less than those required by the standard genetic algorithm. By changing the value of $nrj$, the evolution processes of the structure $P_{cr}$ and $gra\_r_{min}$ are almost the same, and only the final optimization results are slightly different. The optimization process of the guided genetic algorithm is illustrated in detail by taking $nrj = 169$ as an example.

When $nrj = 169$, the optimization process of $P_{cr}$ and $gra\_r_{min}$ is shown in Fig. 3.18. Figure 3.18 shows that, in the first 80 steps of optimization, $P_{cr}$ and $gra\_r_{min}$ increase rapidly and basically reach the optimal value; for the single-layer lattice shell structure with medium span, after the guided genetic algorithm searches to the optimal solution, the structure $P_{cr}$ and $gra\_r_{min}$ remain stable and no longer fluctuate. The stable bearing capacity of the optimized structure is $P_{cr} = 32.55$ kN/m², which is 22.1% higher than that of the initial structure.

Since the structure is symmetric in rotation and load distribution is symmetric, the distribution of bars in the six sectors of the optimized structure is completely the same. Take the black-bolded sector in Fig. 3.13 for illustration, the distribution of bars in the optimized lattice shell structure is shown in Fig. 3.19. The steel consumption of bars is 24.30 kg/m², which is less than 24.46 kg/m² of the initial structure, meeting the constraint condition of steel consumption of bars. In the optimized structure, without human intervention, there are 7 cross-sectional forms of 1122 bars, which is convenient for construction. Under the design load, the final optimized structure meets all the design constraint conditions. The specific values are shown in Table 3.3.

For the optimized lattice shell structure, according to the JGJ7-2010 Technical specification for space grid structure [1], the stability critical load $P_{cr}^{im}$ of the defective structure is $P_{cr}^{im} = 18.39$ kN/m², which is 46.1% higher than that of the initial structure $P_{cr}^{im} = 12.59$ kN/m². For the optimized structure, according to the safety factor method

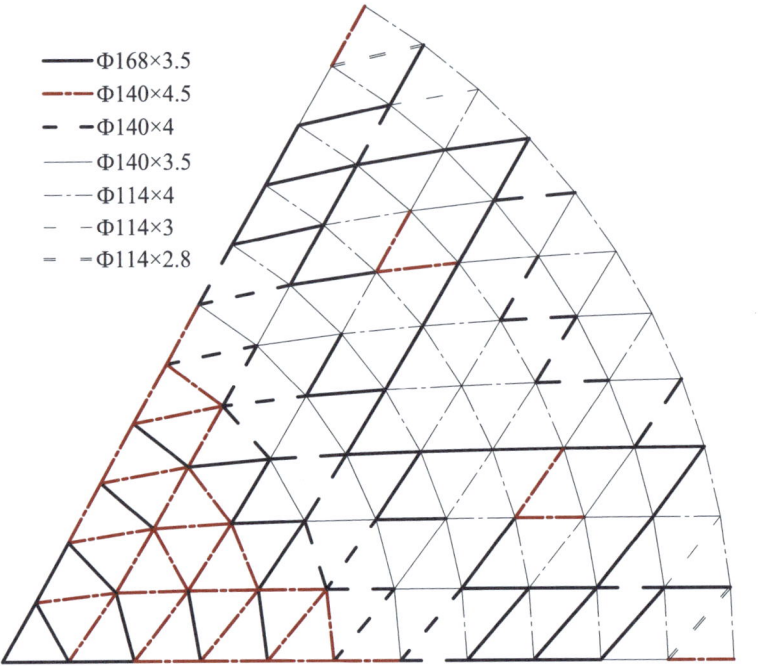

**Fig. 3.19** Sections of 50 m span lattice shell structure after optimization

**Table 3.3** Mechanical indexes of optimized structure of 50 m span reticulated shell

| Mechanical index | Computed value | Limit value |
|---|---|---|
| Maximum stress (N/mm$^2$) | 127.96 | 310 |
| Maximum plane stable stress (N/mm$^2$) | 127.80 | 310 |
| Maximum plane stability stress (N/mm$^2$) | 292.97 | 310 |

in the JGJ7-2010 Technical specification for space grid structure [1], the stability checking is carried out, and $P_{cr}^{im}/4.2 = 4.38$ kN/m$^2$ > 2.55 + 0.50 = 3.05 kN/m$^2$, which passes the stability checking.

### 3.4.4 Comparison of Optimization Algorithms

The calculation results of the standard genetic algorithm and the guided genetic algorithm are shown in Table 3.4. The optimization algorithms are run on the same computer with WIN7 operating system, which is configured with Inter(R) Core(TM) i7-4790 K CPU@ 4.00 GHz and 32 GB memory. It can be concluded from Table 3.4 that: (1) compared with the standard genetic algorithm, the guided genetic algorithm

## 3.4 Optimization Example 2: 50 m Span Single-Layer Reticulated Shell

requires a smaller population size and fewer evolution generations; (2) no matter what the value of the parameter *nrj* in the guided genetic algorithm is, the optimization process of $P_{cr}$ and $gra\_r_{min}$ is basically the same, and the $P_{cr}$ of the final optimized lattice shell is basically the same, which indicates that the guided genetic algorithm has low parameter sensitivity and good robustness; (3) because the standard genetic algorithm adopts random mutation, the cross-section types in the optimization results are too many, which is difficult to meet the actual construction needs; (4) for the lattice shell structure, the time required by the guided genetic algorithm is about 70 min, which is significantly less than the standard genetic algorithm.

The evolution of structural $P_{cr}$ under different optimization parameters of the two algorithms is shown in Fig. 3.20. The random mutation mechanism in the standard genetic algorithm cannot guarantee the continuity of individuals before and after mutation, so the optimization process of structural $P_{cr}$ is unstable with significant oscillations. Meanwhile, for the single-layer reticulated shell structure with medium span, even if the population size is 100, the standard genetic algorithm cannot improve the structural stability bearing capacity. Instead of the standard genetic algorithm, the guided genetic algorithm can ensure that the final optimization results converge to the optimal solution. In the guided genetic algorithm, the difference of the optimization parameter *nrj* will lead to the difference of the evolution rate of $P_{cr}$. However, no matter what the value of *nrj* is, $P_{cr}$ basically remains stable after 60 optimization steps. At the same time, no matter what the value of *nrj* is, the $P_{cr}$ value of the optimization results is basically the same.

**Table 3.4** Comparison of optimization results of 50 m span gridshell

| Calculation parameters | GA | GA | GGA | | |
|---|---|---|---|---|---|
| | | | $nrj = 91$ | $nrj = 127$ | $nrj = 169$ |
| Population size | 5 | 100 | 5 | 5 | 5 |
| Evolutionary algebra | 400 | 400 | 150 | 150 | 150 |
| $P_{cr}$ (kN/m$^2$) | 27.05 | 28.44 | 30.65 | 31.88 | 32.55 |
| Amount of steel used (kg/m$^2$) | 24.04 | 24.07 | 24.16 | 24.11 | 24.01 |
| $gra\_r_{min}$ ($10^{-6}$) | $-4.148$ | $-3.520$ | $-3.668$ | $-3.668$ | $-3.576$ |
| Type of section | 37 | 30 | 7 | 7 | 7 |
| Time consuming (min) | 104 | 3166 | 40 | 44 | 73 |
| Time consuming (min/generation) | 0.26 | 7.915 | 0.2667 | 0.2933 | 0.4867 |

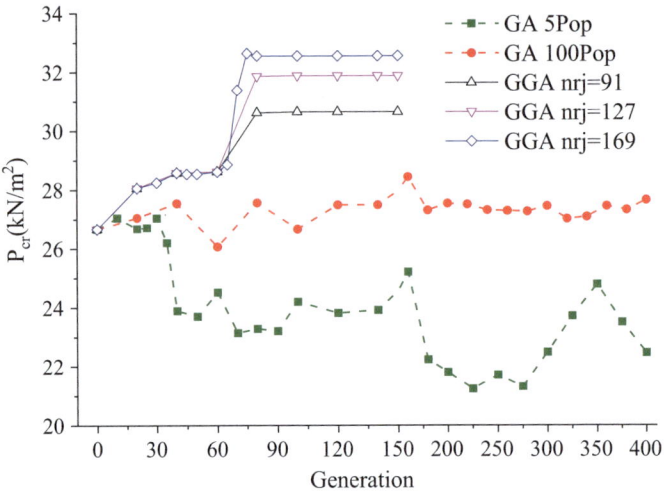

**Fig. 3.20** Optimization history of each optimization algorithm $P_{cr}$

## 3.5 Optimization Example 3: 80 m Span Single-Layer Reticulated Shell

### 3.5.1 Basic Information of Structure

As shown in Fig. 3.21, the long-span K6 single-layer lattice shell has a span of 80 m, a beam height of 25 m, and fixed supports around it. The structure has a total of 2352 bars and 817 joints. The lattice shell structure is also composed of 6 identical sectors, and a sector is shown as the black bold bars in Fig. 3.21.

The structural steel is Q345. The standard value of dead load is 2.20 kN/m², the standard value of live load is 0.50 kN/m², and the design value of load is 3.46 kN/m². The optimization variable is the section of the rod, and the value range is the section with the outer diameter of 89–406 mm and the wall thickness not greater than 20 mm in GB/T 17395-2008 Seamless Steel Tube Dimensions, Shape, Weight and Allowable Deviation [2], a total of 196 candidate sections. The candidate sections almost include all suitable sections. The upper limit of the steel amount of the rod is set as 46 kg/m². After full stress design, the steel amount of the initial structure is 45.66 kg/m², which is basically equal to the limit of the steel amount, thus ensuring that the steel amount of the rod is basically equal in the optimization process.

3.5 Optimization Example 3: 80 m Span Single-Layer Reticulated Shell

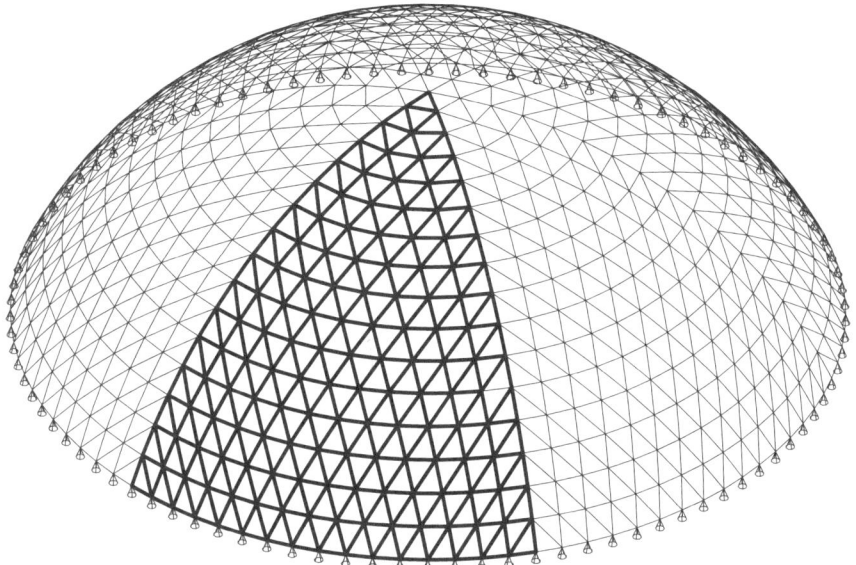

**Fig. 3.21** 80 m span single layer spherical reticulated shell

### 3.5.2 Guided Genetic Algorithm for Solving Stable Optimization

It can be seen from Sect. 3.4 that for the single-layer lattice shell structure with medium span, the standard genetic algorithm cannot search for the optimal solution within an acceptable time. Therefore, for the large-span K6 single-layer lattice shell shown in Fig. 3.21, the guided genetic algorithm is directly adopted for optimization. The load mode is the controlling load mode of single-layer spherical lattice shell—full span uniform load mode. In the improved optimization algorithm, the number of weak joints is $nvj = 7$, the crossover rate is still 0.2, the size of population is 5, and the number of redundant joints is $nvj = 169$. See Fig. 3.22 for the evolution process of structure $P_{cr}$ and $gra\_r_{min}$.

It can be seen from Fig. 3.22 that, under the premise of given upper limit of steel consumption, the improved optimization algorithm can gradually improve the structural stable bearing capacity and converge to the optimal solution. According to the nonlinear tracking of arc length method, the elastic stable bearing capacity of the initial structure is $P_{cr} = 30.67$ kN/m². The elastic stable bearing capacity of the optimized structure is $P_{cr} = 40.01$ kN/m², $gra\_r_{min} = -7.9242 \times 10^{-6}$. Compared with the initial structure, the elastic stable bearing capacity of the optimized structure is increased by 30.5%. The steel consumption of the optimized structure is 45.34 kg/m² < 46 kg/m², meeting the constraint conditions of steel consumption.

Since the structure is symmetric in rotation and load distribution is also symmetric, the distribution of bars in the six sectors of the optimized structure is completely the

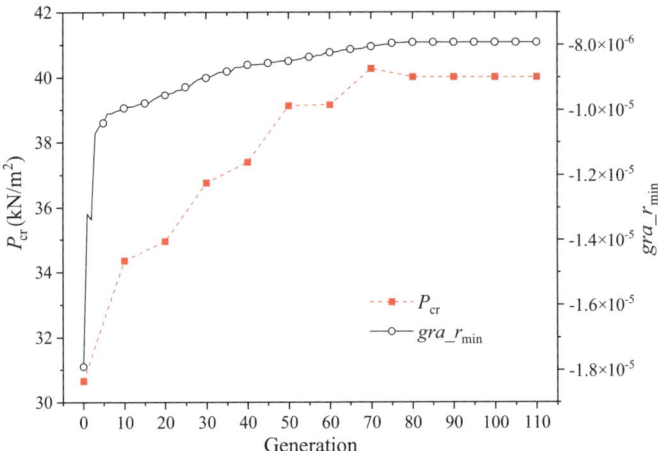

**Fig. 3.22** Evolution of $P_{cr}$ and $gra\_r_{min}$ in 80 m span single-layer reticulated shells

same. Take the black-bold sector in Fig. 3.21 for illustration, and the distribution of bars after optimization is shown in Fig. 3.23. Under the design load, the optimized structure meets all mechanical constraint conditions, and the specific values are shown in Table 3.5.

For the optimized lattice shell structure, the stability critical load $P_{cr}^{im}$ of the defective structure is $P_{cr}^{im} = 12.14$ kN/m² according to the JGJ7-2010 Technical Specification for Space Grid Structures [1]. For the optimized structure, the stability checking is carried out according to the safety factor method in the JGJ7-2010 Technical Specification for Space Grid Structures [1] (the safety factor is 4.2). The stability checking is passed when $P_{cr}^{im}/4.2 = 2.89$ kN/m² $> 2.20 + 0.50 = 2.70$ kN/m².

## 3.6 Chapter Summary

Based on the instability mechanism of lattice shell structure, this chapter proposes the stability optimization design method of single-layer lattice shell structure, and develops the corresponding efficient optimization algorithm. In the optimization model, from the opposite side of the structure to maintain stability, that is, from the angle of the structure loss of stability, the simple $gra\_r_{min}$ is calculated to represent the degree of structural stiffness degradation and measure the structural stability bearing capacity. The optimization goal is to reduce the degree of structural softening, so as to improve the structural stability bearing capacity. The optimization variable is the section of the rod. To consider the actual construction requirements, the section of the rod is taken from the Chinese manufacturing standard GB/T 17395-2008 Seamless Steel Tube Size, Shape, Weight and Allowable Deviation [2], which is a discrete variable. The mandatory design requirements stipulated in the Chinese

## 3.6 Chapter Summary

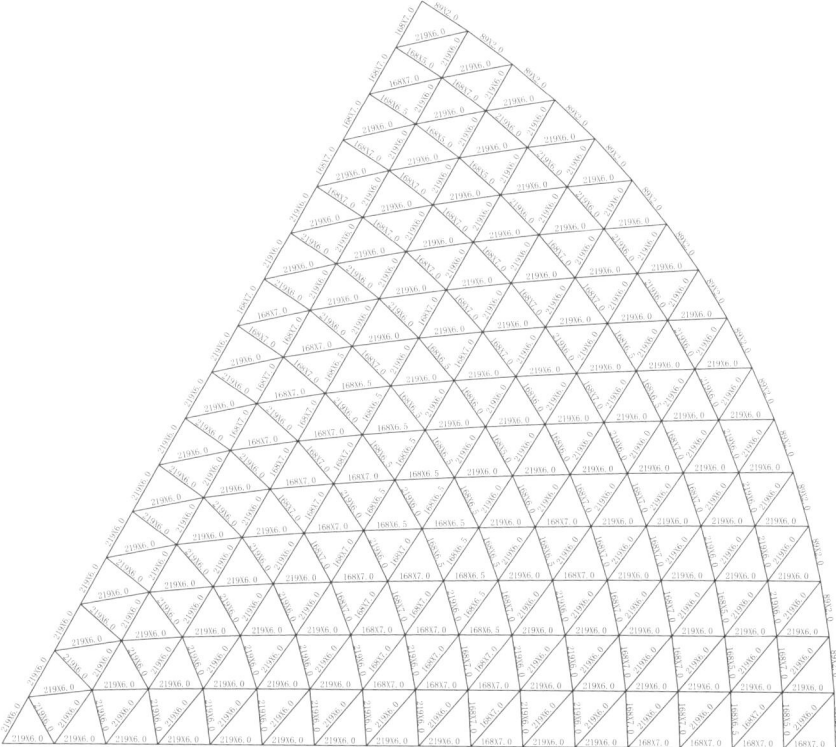

**Fig. 3.23** Sections of 80 m span lattice shell structure after optimization (section outer diameter × thickness, Unit: mm)

**Table 3.5** Mechanical indexes of optimized structure of 80 m span reticulated shell

| Mechanical index | Computed value | Limit value |
|---|---|---|
| Maximum stress (N/mm$^2$) | 66.60 | 310 |
| Maximum plane stable stress (N/mm$^2$) | 66.46 | 310 |
| Maximum plane stability stress (N/mm$^2$) | 98.50 | 310 |

design specifications are the optimization constraint conditions. The optimization model can carry out the anti-instability optimization design of single-layer lattice shell structure under the premise of a given upper limit of the steel amount of the rod.

In view of the defects of the large number of variables in the stable optimization model, the low efficiency of the standard genetic algorithm and the continuity cannot be satisfied, this chapter improves the random mutation mechanism in the standard genetic algorithm and puts forward the guided genetic algorithm. The guided genetic algorithm can carry out directional mutation with high search efficiency, and can quickly realize the stable optimization of large single-layer lattice shell structure. The final result converges to the optimal solution and obtains the stable and reliable optimization design result. Finally, taking a 22 m small span lattice shell, a 50 m medium span lattice shell and an 80 m long span lattice shell as examples, the stable optimization design is carried out under the premise of the economy of steel consumption. In the stable optimization process of the 22 m small span lattice shell, the standard genetic algorithm takes 780 min to obtain the optimization result of $P_{cr} = 93.84$ kN/m$^2$, while the guided genetic algorithm only takes 11 min to obtain the optimization result of $P_{cr} = 94.36$ kN/m$^2$, and the steel consumption remains unchanged. In the stable optimization of the 50 m medium span lattice shell structure, the conventional design of the lattice shell has $P_{cr} = 26.66$ kN/m$^2$, and the standard genetic algorithm takes 3166 min to obtain the optimization result of $P_{cr} = 28.44$ kN/m$^2$, and the optimization result does not converge. The guided genetic algorithm takes 73 min to obtain the optimization result of $P_{cr} = 32.55$ kN/m$^2$. Under the premise of the same steel consumption, the structure obtained by the guided genetic algorithm has a 22.1% higher $P_{cr}$ than the conventional design of the lattice shell; compared with the structure obtained by the standard genetic algorithm, the $P_{cr}$ has increased by 14.5%, and the optimization results are stable and reliable. In the stability optimization of 80 m long span single layer lattice shell structure, the standard genetic algorithm has been unable to solve such a large-scale structural stability optimization problem, while the guided genetic algorithm takes 110 min to get the optimization result of $P_{cr} = 40.01$ kN/m$^2$, and the optimization results are stable and reliable.

# References

1. Industrial standard of the People's Republic of China (2010) JGJ7-2010 technical specification for space grid structure. China Architecture and Building Press, Beijing
2. National Standard of the People's Republic of China (2008) GB/T 17395–2008 dimension, shape, weight and allowable deviation of seamless steel pipes. Standards Press of China, Beijing
3. Kicinger R, Arciszewski T, De Jong K (2005) Evolutionary computation and structural design: a survey of the state-of-the-art. Comput Struct 83(23–24):1943–1978
4. Adeli H, Cheng NT (1994) Concurrent genetic algorithms for optimization of large structures. J Aerosp Eng 7(3):276–296
5. National Standard of the People's Republic of China (2003) GB 50017–2003 code for design of steel structures. China Architecture and Building Press, Beijing
6. Gen M, Cheng R (1996) A survey of penalty techniques in genetic algorithms. In: Proceedings of IEEE international conference on evolutionary computation. IEEE, Nagoya, Japan, pp 804–809
7. Wang X, Cao L (2002) Genetic algorithms-theory, application and software implementation. Xi'an Jiaotong University Press, Xi'an, pp 5–20

**Open Access** This chapter is licensed under the terms of the Creative Commons Attribution-NonCommercial-NoDerivatives 4.0 International License (http://creativecommons.org/licenses/by-nc-nd/4.0/), which permits any noncommercial use, sharing, distribution and reproduction in any medium or format, as long as you give appropriate credit to the original author(s) and the source, provide a link to the Creative Commons license and indicate if you modified the licensed material. You do not have permission under this license to share adapted material derived from this chapter or parts of it.

The images or other third party material in this chapter are included in the chapter's Creative Commons license, unless indicated otherwise in a credit line to the material. If material is not included in the chapter's Creative Commons license and your intended use is not permitted by statutory regulation or exceeds the permitted use, you will need to obtain permission directly from the copyright holder.

# Chapter 4
# Stability Optimization and Collapse Resistance Verification of Large Single-Layer Lattice Shell Structure

**Abstract** To further verify the applicability of the stability optimization design method, this chapter takes two practically constructed large single-layer lattice shell structures used for shaking table collapse experiments as examples to carry out stability optimization design. Each lattice shell structure has as many as 3660 bars, with a very large number of optimization variables. The candidate bar section ranges from outer diameter 18 mm to outer diameter 114 mm, with a similarly large number of candidate bars. The support conditions, topological connections and steel consumption of the two lattice shell structures are completely the same, but Model 1 is a perfect structure with uniform stiffness, while Model 2 is a weak model with artificially set weak areas and stiffness redundancy areas. The two practical models are optimized and the optimization results are compared to further verify the stability optimization design method. The actual built models 1 and 2 were subjected to shaking table collapse experiments. In order to verify the collapse resistance ability of the stabilized and optimized structure, the same seismic wave was input into the two optimized lattice shell structures through numerical simulation, and the display dynamic algorithm was used to study the collapse resistance ability of the stabilized and optimized structure. The checking results show that the stabilized and optimized structure not only has good static stability performance, but also has good collapse resistance performance.

## 4.1 Large-Span Single-Layer Gridshells

### 4.1.1 Model Structure Design

The span of the two K6 single-layer spherical reticulated shell test models is 81.9 m, the height-to-span ratio is 0.5, and the connection topology of the bars is completely the same. Because the space grid structure has the characteristics that the strength and stiffness cannot be similar to the prototype at the same time under small scale ratio, the scale ratio of this test is 1:3.5, the model span is 23.4 m, and the height of the vector is 11.7 m. The test model is strictly in accordance with the topology of the

**Table 4.1** Model and prototype size

| Item | Prototype | Model |
|---|---|---|
| Span | 81.9 m | 23.4 m |
| Height | 40.95 m | 11.7 m |
| Vector span ratio | 0.5 | 0.5 |

prototype. See Table 4.1 for the model size and the prototype size. The total number of joints (welded hollow spheres) in each model is 1261, and the total number of bars is 3660. On the basis of meeting the requirements of geometric similarity, the model also meets the requirements of load similarity, mass similarity, and stiffness similarity and so on.

Model design purpose: Model 1 adopts full stress design with uniform overall stiffness, so as to observe the whole process of the structure from elasticity to plasticity until collapse; Model 2 has uneven overall stiffness, and two weak areas are set in the region with large dynamic response, and the section of the top bar is increased, so that the steel amount of the two models is the same, so as to obtain the difference in the collapse process with Model 1 [1, 2].

The standard load values of the prototype structure are as follows: dead weight 0.35 kN/m$^2$, roof slab dead weight 0.55 kN/m$^2$, and live load 0.50 kN/m$^2$. Through the similarity ratio design and full stress design, the section specifications of the bar parts of Model 1 are $\Phi 23 \times 1$, $\Phi 38 \times 2$, $\Phi 63.5 \times 3.5$, and $\Phi 114 \times 4$; the weak area bar part of Model 2 is $\Phi 18 \times 1$, the top reinforcement bar part is $\Phi 38 \times 2$, and other bar parts are the same as Model 1. The constitutive relation of materials is set as Q235 steel. Because of the symmetrical characteristics of the two test models, the 1/4 structure is taken to represent the distribution of the bar parts, as shown in Figs. 4.1 and 4.2, respectively. The actual structure model is shown in Fig. 4.3. The spatial position of the weak area in the whole structure of Model 2 can be seen in Fig. 4.4.

The weight of the model consists of two parts: the mass corresponding to the representative value of gravity load and the additional mass of the scale model to meet the gravity density similarity relationship, the specific value should be derived according to the static design load, the similarity relationship between the prototype and the model [1]. The final weight of each joint of Model 1 was determined to be 30 kg. Based on a similar derivation, the weight of Model 2 was determined to be 20 kg for each joint in the weak area and 30 kg for other joints.

The test was completed by using the multi-point shaking table test system of Tongji University. See Fig. 4.3 for the actual structural model and shaking table. The system consists of four shaking tables: A (30 tons side table), B (70 tons main table), C (70 tons main table) and D (30 tons side table). See Fig. 4.4 for the location and number of each shaking table. There are 40 supports around the test model, 10 supports on each shaking table (see Fig. 4.5), and no supports are set around other joints.

## 4.1 Large-Span Single-Layer Gridshells

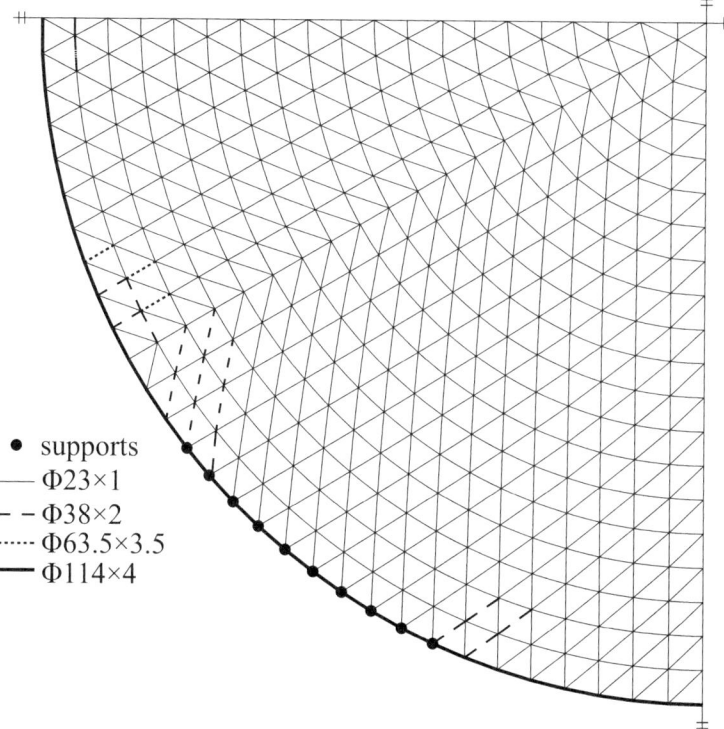

**Fig. 4.1** Model 1 initial section designation

### 4.1.2 Model Shaking Table Collapse Test

The input seismic wave took EL-Centro seismic wave (1940) as reference. Each shaking table input an analog seismic wave, with the input direction along the Y direction (see Fig. 4.4). The seismic wave itself met the set self-spectrum relationship, and different seismic waves met the set cross-spectrum relationship. Considering the similarity relationship, the seismic wave duration was 28.58506 s, and the time interval was 0.01069 s. The seismic wave input by each shaking table was shown in Fig. 4.6 [1]. In the test process, the peak value of seismic acceleration (PGA) was gradually increased according to the conditions in Table 4.2, and the collapse and failure process of the two models were observed respectively.

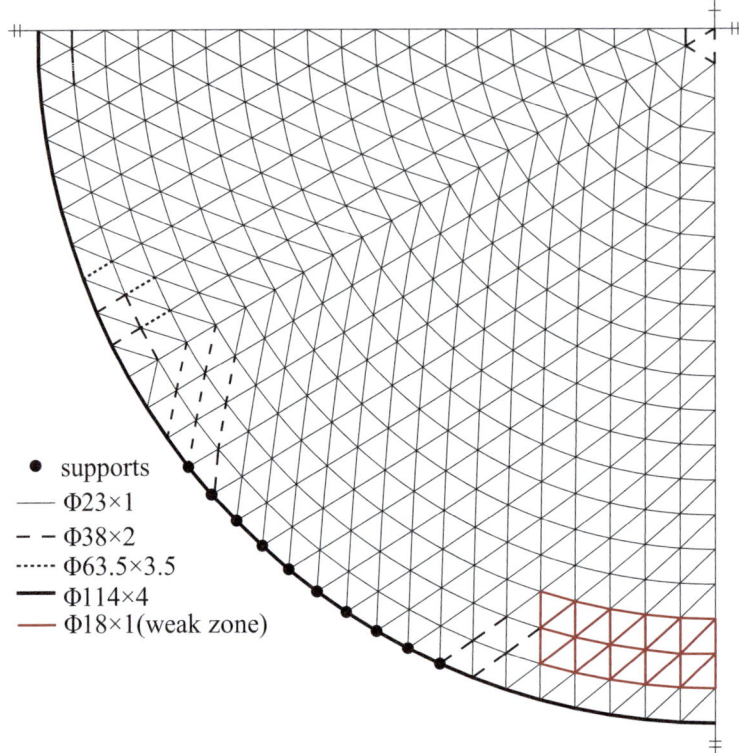

**Fig. 4.2** Model 2 initial section designation

**Fig. 4.3** Actual model and shaking tables

4.1 Large-Span Single-Layer Gridshells

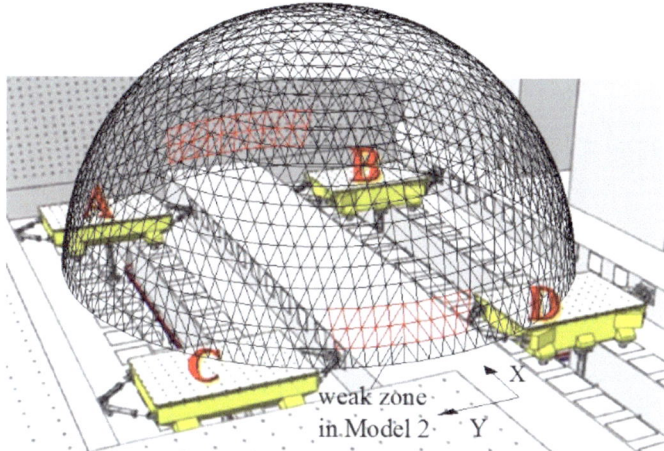

**Fig. 4.4** Shaking tables and reticulated shell model

**Fig. 4.5** Constraints of the models

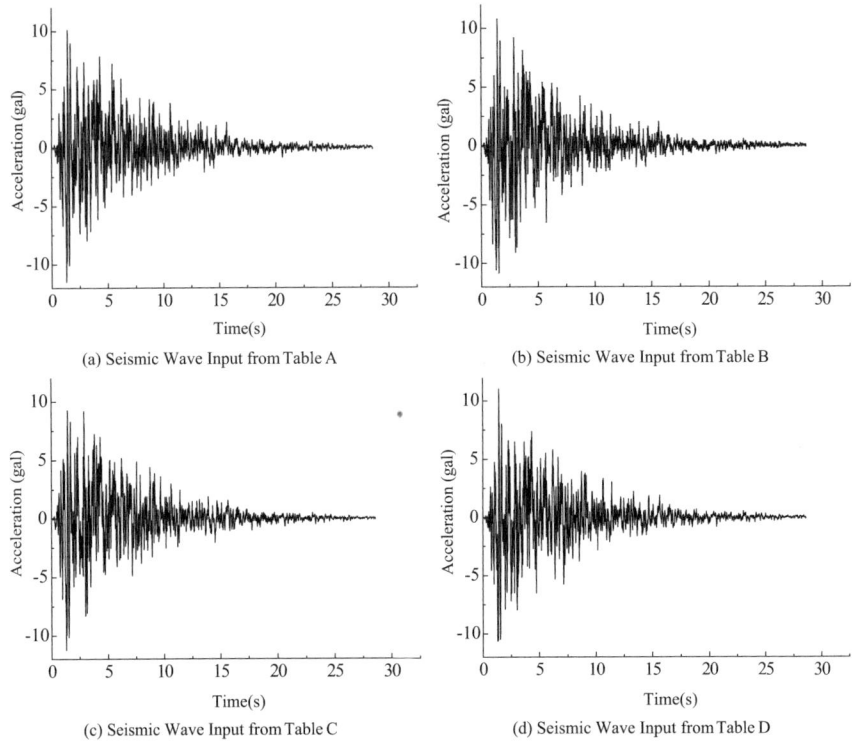

**Fig. 4.6** Shaking table test input seismic waveform

**Table 4.2** Test loading condition

| Operating condition | Model 1 | Model 2 |
|---|---|---|
| 1 | First white noise sweeping | First white noise sweeping |
| 2 | 100 gal | 100 gal |
| 3 | 200 gal | Second white noise sweeping |
| 4 | Second white noise sweeping | 200 gal |
| 5 | 250 gal | 250 gal |
| 6 | 300 gal | Third white noise sweeping |
| 7 | Third white noise sweeping | 300 gal |
| 8 | 350 gal | Fourth white noise sweeping |
| 9 | 400 gal | 350 gal |

## 4.2 Stability Optimization of Large Single-Layer Lattice Shell Structure

### 4.2.1 Model 1 (Normal Model)

According to the calculation, in the initial structure of Model 1, the minimum value of relative change gradient of joint well-formedness $gra\_r_{min} = -5.457 \times 10^{-4}$, the initial steel amount $V_0 = 0.35$ m$^3$, and the critical load of structural stability $P_{cr} = 3.27$ kN/m$^2$ obtained by arc length tracking method. Combined with the section specification of the initial structure bar, the value range of the optimization variable (i.e. the section of the bar) is all the bars with an outer diameter of 18–114 mm and a wall thickness not greater than 4 mm in GB/T 17395-2008 Seamless Steel Tube Dimensions, Configuration, Weight and Allowable Deviation [3], and there are 575 candidate section types. The value range of the candidate section almost includes all the suitable bars. Because the bearing is discontinuous and there is stress concentration in the model, the constraint condition of steel amount is slightly relaxed, and the adjustment coefficient of steel amount $\eta_v = 1.07$ is taken.

In the whole optimization process, the optimization process of $gra\_r_{min}$ and $P_{cr}$ is shown in Figs. 4.7 and 4.8 respectively. Take the projection position of each joint on the horizontal plane as X and Y coordinates, and $gra\_r$ as Z coordinates, draw the distribution cloud map of $gra\_r$, and the optimization process of $gra\_r$ distribution is shown in Fig. 4.9.

Combined with Figs. 4.7, 4.8, and 4.9: $gra\_r_{min}$ increases most obviously in the first 70 optimization steps, $gra\_r$ distribution rapidly tends to be uniform, and correspondingly, $P_{cr}$ also increases significantly in this stage; after 70 steps, $gra\_r_{min}$ converges to a constant value and remains stable, while $P_{cr}$ only slightly increases; when the evolution reaches 210 generations, both $gra\_r_{min}$ and $P_{cr}$ have reached the

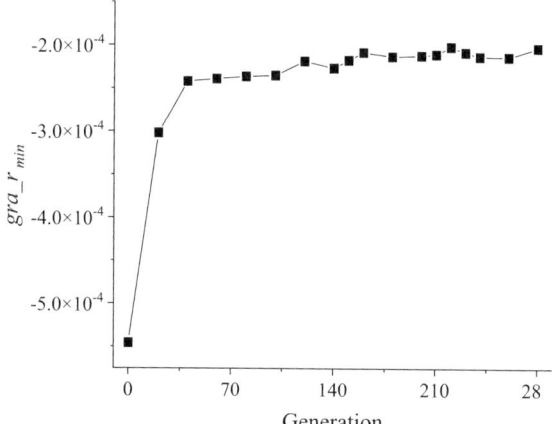

**Fig. 4.7** The $gra\_r_{min}$ optimization process of model 1

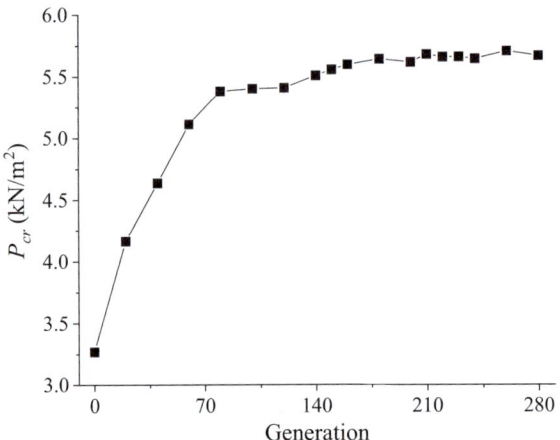

**Fig. 4.8** The $P_{cr}$ optimization process of model 1

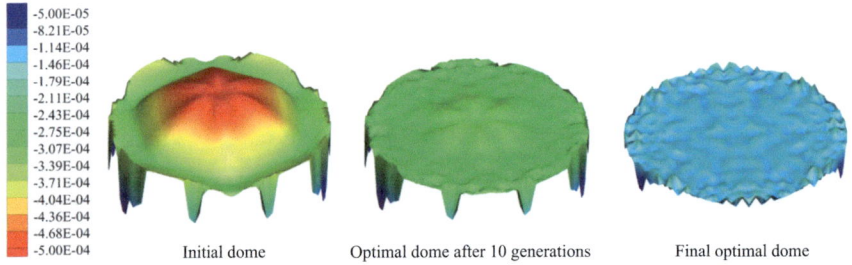

**Fig. 4.9** Optimization process of *gra_r* distribution of model 1

optimal value; when the evolution continues to 280 generations, both $gra\_r_{min}$ and $P_{cr}$ can remain stable until the program terminates.

After stable optimization, the distribution of the structural members is shown in Fig. 4.10, and the section specifications of the members are marked on the top of the members in the form of diameter (mm) × wall thickness (mm). K6 lattice shell has 6 identical sectors, and 40 fixed supports are symmetrically distributed around the structure. Under the uniform load of full span, the members of K6 lattice shell after stable optimization are also symmetrically distributed. Therefore, in Fig. 4.10, a sector without support and a sector with support respectively represent the distribution of the members of the overall structure.

The optimized structure $gra\_r_{min} = -2.029 \times 10^{-4}$ is effectively improved compared with the initial structure, indicating that the maximum softening degree of the structure is significantly reduced; *gra_r* is evenly distributed, indicating that the softening degree of each part is basically the same, the overall anti-load potential of the structure is maximized, and the overall structure is reasonably stressed; $V = 0.3724$ m³ $< \eta_v \times V_0$, meeting the constraint conditions of steel quantity; $P_{cr} =$

5.67 kN/m², 1.732 times higher than the initial structure; the instability deformation is magnified by 5 times as shown in Fig. 4.11. Because the structure is supported on 4 shaking tables, the bars near the support are obviously stressed, resulting in the optimized structure only buckling the bars above the support, and there is no obvious buckling deformation in other parts.

### 4.2.2 Model 2 (Weak Model)

There are two symmetric weak stiffness areas in Model 2, and the positions of the weak areas in the structure are shown in Figs. 4.2 and 4.4. Meanwhile, the first circle

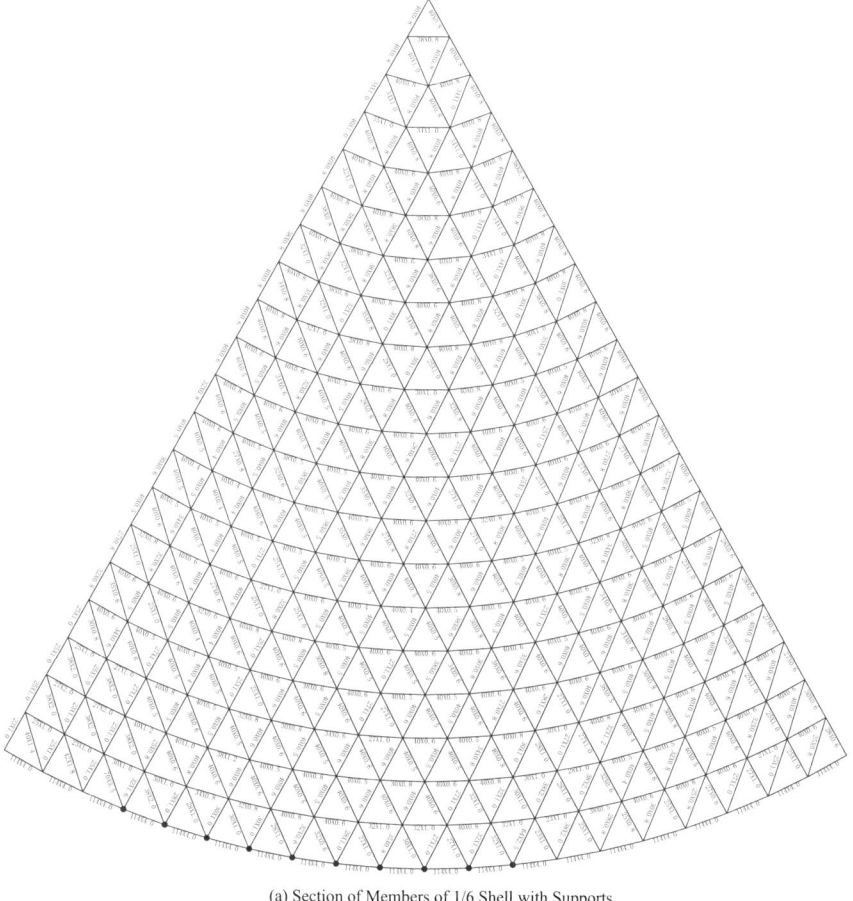

(a) Section of Members of 1/6 Shell with Supports

**Fig. 4.10** Sestion of members after optimization of model 1

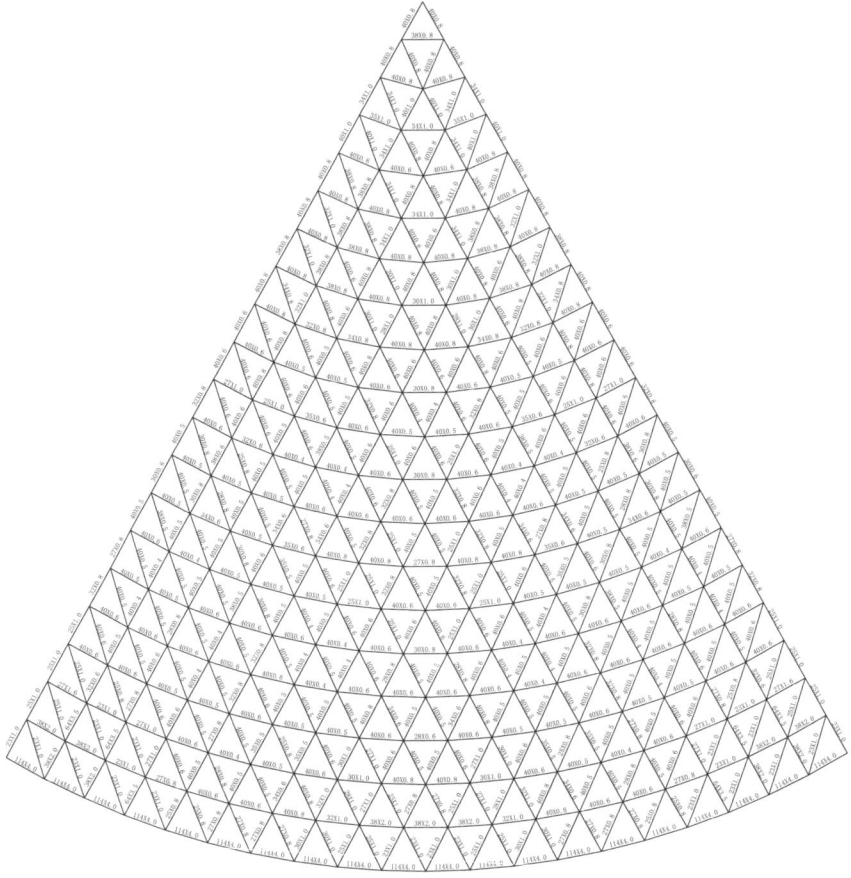

(b) Section of Members of 1/6 Shell without Supports

**Fig. 4.10** (continued)

of inclined bar at the top has a high stiffness, and the stiffness distribution of the whole structure is uneven and irregular. Through calculation, the initial structure of Model 2 uses $V_0 = 0.3498$ m$^3$ of steel, and the minimum value of relative change gradient of joint well-formedness $gra\_r_{\min} = -6.9908 \times 10^{-4}$; in the vertex strengthening area, the vertex $gra\_r = -7.0374 \times 10^{-5}$ is one order of magnitude higher than the surrounding joints. The stable tracking by arc length method shows that the stable bearing capacity of the initial structure is $P_{cr} = 3.13$ kN/m$^2$. Model 2 is optimized stably, and the value range of optimization variables is the same as that of Model 1. The optimization process of $gra\_r_{\min}$ and $P_{cr}$ is shown in Figs. 4.12 and 4.13 respectively, and the optimization process of $gra\_r$ distribution is shown in Fig. 4.14.

In combination with Figs. 4.12, 4.13, and 4.14: the distribution of $gra\_r$ in the initial structure is extremely irregular, and its minimum value $gra\_r_{\min} = -6.9908 \times 10^{-4}$ is located in the weak area, and $gra\_r$ in the vertex strengthening area is

## 4.2 Stability Optimization of Large Single-Layer Lattice Shell Structure

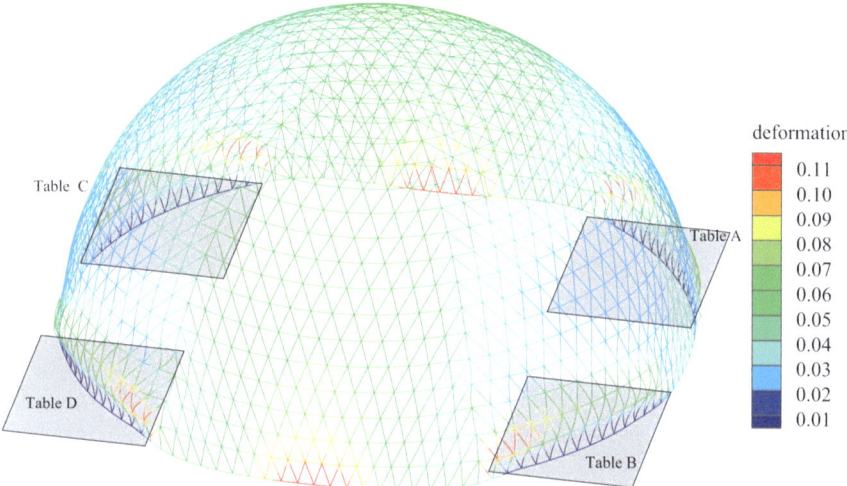

**Fig. 4.11** Structural instability deformation after optimization of model 1 (unit: m)

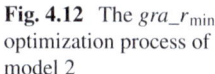

**Fig. 4.12** The $gra\_r_{\min}$ optimization process of model 2

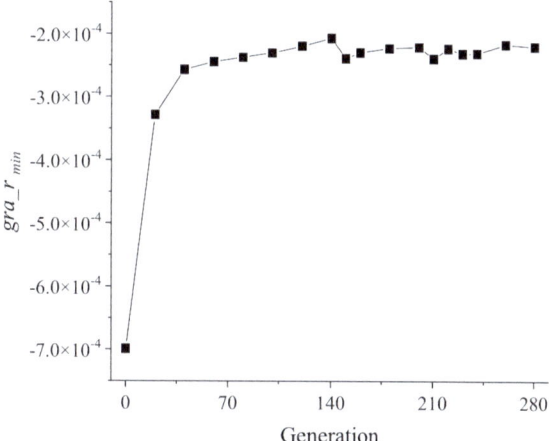

significantly higher than that of the surrounding joints, indicating that the overall stiffness distribution is unreasonable; in the optimization process, $gra\_r_{\min}$ gradually increases, and the overall softening degree of the structure gradually decreases; at the same time, the distribution of $gra\_r$ in the weak area tends to be uniform, indicating that the optimization algorithm can accurately identify the weak area and moderately strengthen the weak area in the directional mutation; the distribution of $gra\_r$ near the vertex also gradually becomes smooth, indicating that the optimization algorithm can accurately identify the stiffness redundancy area and weaken the bars in the stiffness redundancy area; with the decrease of the degree of structural softening (the increase of $gra\_r_{\min}$) and the rationalization of the structural stiffness distribution

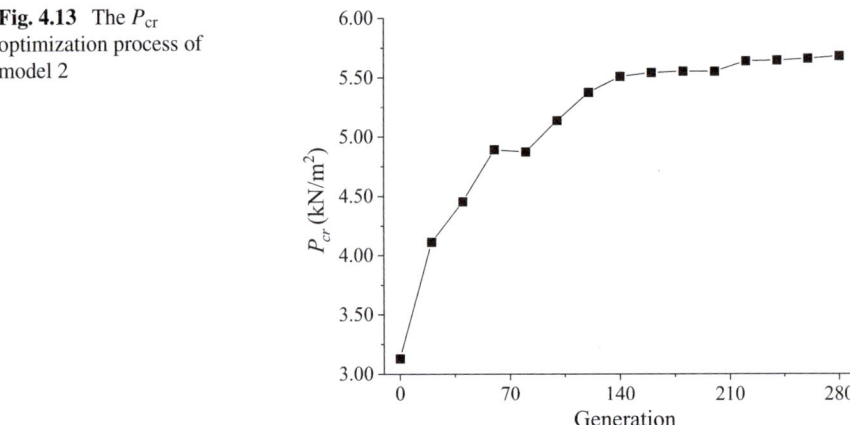

**Fig. 4.13** The $P_{cr}$ optimization process of model 2

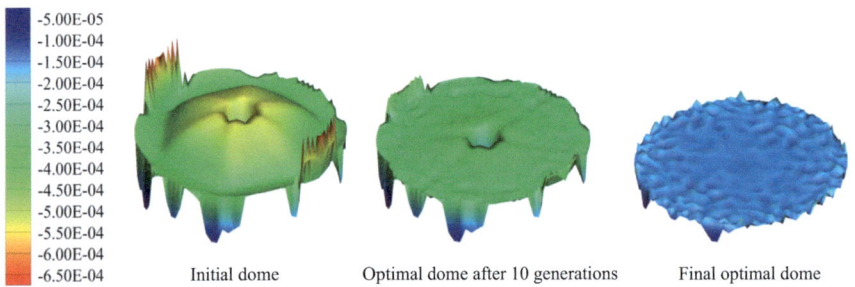

**Fig. 4.14** Optimization process of gra_r distribution of model 2

(the homogenization of gra_r distribution), $P_{cr}$ also gradually increases; after 210 optimization steps, gra_$r_{min}$ and Pcr converge to the optimal value and remain stable; when the evolution reaches 280 generations, gra_r is distributed evenly, and there is no stiffness redundancy area or stiffness weak area, and the optimization is completed.

After stable optimization, the distribution of the rod of Model 2 is shown in Fig. 4.15, and the section specification of the rod is marked on the top of the rod in the form of diameter (mm) × wall thickness (mm). K6 lattice shell has 6 same sectors, and 40 fixed supports are symmetrically distributed around the structure. Under the full span uniform load, the rod of K6 lattice shell after stable optimization is also symmetrically distributed. Therefore, a sector without support and a sector with support are shown in Fig. 4.15 to represent the rod distribution of the overall structure. The position of the weak part in the original structure of Model 2 is the red rod in Fig. 4.15, and the original rod specifications are round steel pipes with an outer diameter of 18 mm and a wall thickness of 1 mm. Without human intervention, after stable optimization, these originally weak rods are strengthened and maintain consistent stiffness with the surrounding rods. At the same time, the original rod in the redundant stiffness area on the top is reasonably weakened without human intervention, and maintains continuous stiffness with the surrounding rods.

The optimized structure $gra\_r_{min} = -2.2040 \times 10^{-4}$, which is effectively improved compared with the initial structure; $gra\_r$ is evenly distributed, indicating that the structure stiffness distribution is reasonable and the overall anti-load potential has been mined to the greatest extent; $V = 0.3741 \text{ m}^3 < \eta_v \times V_0$, meeting the steel quantity constraint condition; $P_{cr} = 5.672 \text{ kN/m}^2$, which is 1.812 times higher than the initial structure.

The initial structure and the optimized structure of Model 2 were tracked by arc length method to obtain the structural instability mode. Because the structural deformation is symmetrical, the semi-structure was selected and enlarged by 5 times as shown in Fig. 4.16. In Fig. 4.16, the red bar is the bar in the weak area of the initial

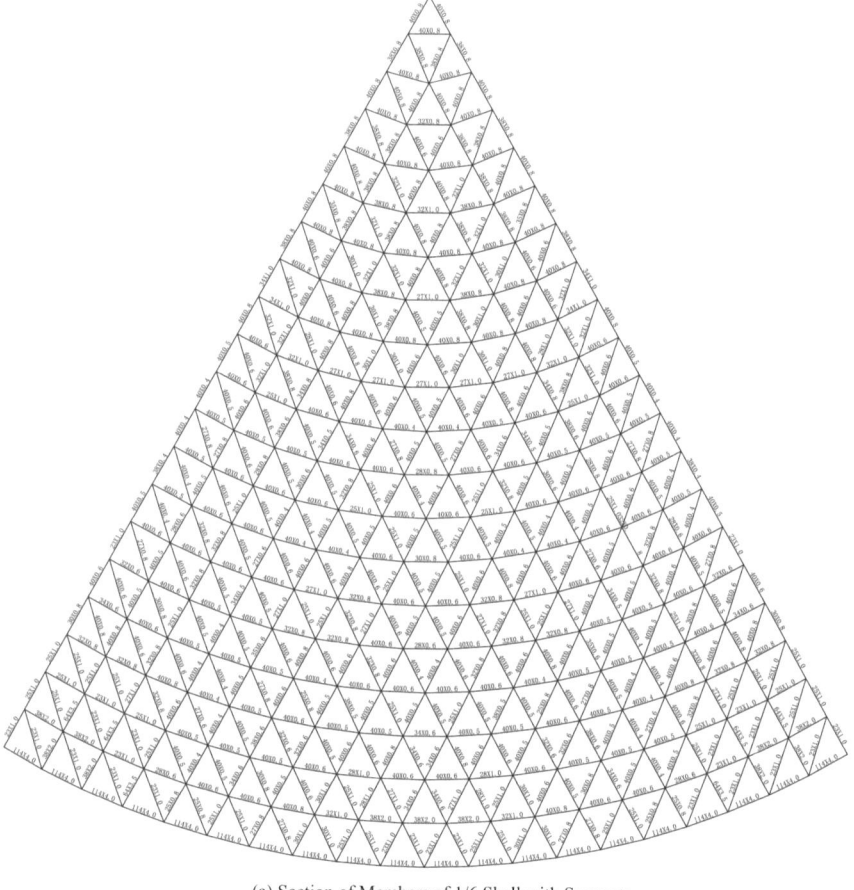

(a) Section of Members of 1/6 Shell with Supports

**Fig. 4.15** Section of members after optimization of model 2

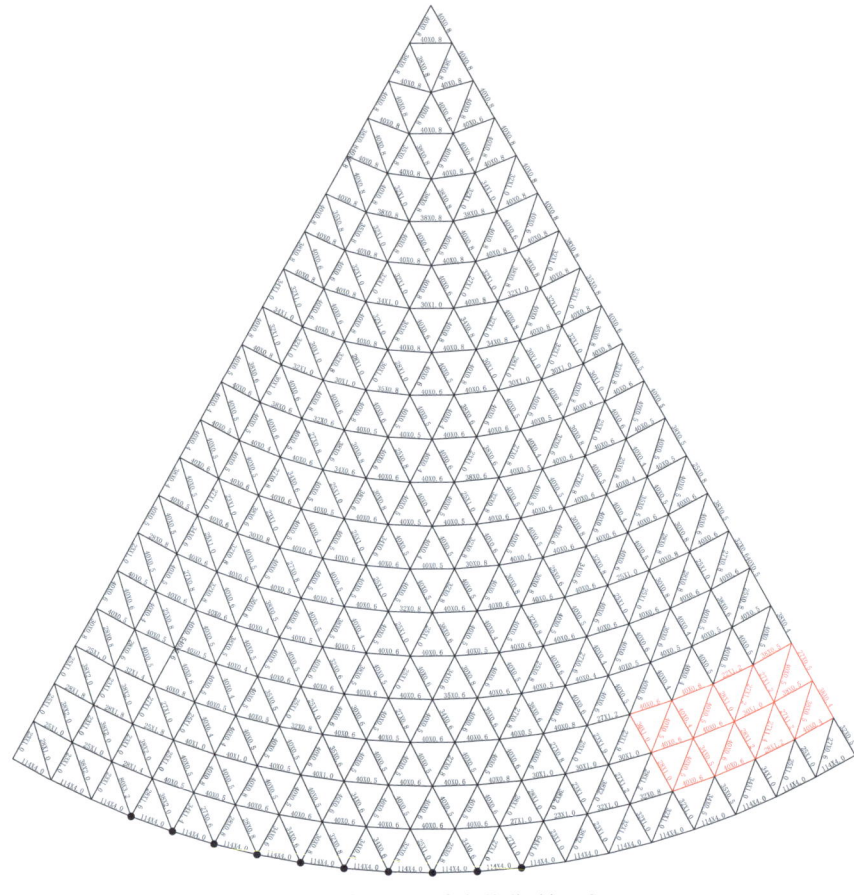

(b) Section of Members of 1/6 Shell without Supports

**Fig. 4.15** (continued)

structure. The distribution of *gra_r* in the initial structure in Fig. 4.14 shows that *gra_r* in the weak area is significantly lower than that in other joints. In the process of stable tracking, the weak area is the first to fail, while the bar above the bearing where the force is concentrated remains stable. The initial structure presents a local instability mode with only two weak areas failing (the red area in Fig. 4.16). In the optimized structure, no obvious instability deformation is seen in the weak area of the initial structure, but there is an instability area above the bearing where the force is concentrated. Besides, the deformation is most significant near the main rib area, which is very similar to the instability deformation of the stable optimized structure of Model 1 (see Fig. 4.11). After optimization, the structure changes from the local instability of two weak areas to the local instability mode of four bearing areas, and the instability mode is improved.

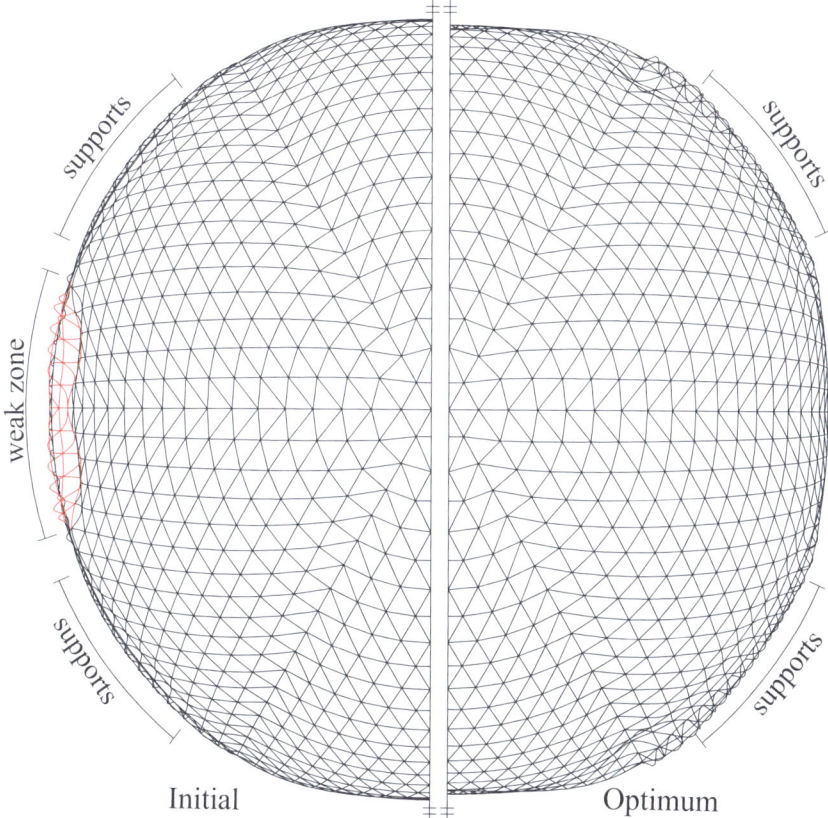

**Fig. 4.16** Model 2 initial structure and instability deformation of optimized structure (top view)

## 4.2.3 Comparison of Stable Optimization of Two Models

The comparison of stable optimization results of the two models under uniformly distributed loads is shown in Table 4.3, and the structural instability modes after optimization are shown in Figs. 4.11 and 4.16 respectively. According to Table 4.3, it can be concluded that: (1) for the two large-scale actual lattice shell structures, the stable bearing capacity of the lattice shell structure can be effectively improved by maximizing $gra\_r_{min}$ and reducing the degree of structural stiffness degradation, which verifies the correctness of the instability mechanism analysis method of the lattice shell structure and the correctness of the optimization model. (2) Under the premise of the same structure topological connection, the same bearing constraints and the same steel amount, after optimization, the two optimized structures have basically the same steel amount, similar $gra\_r_{min}$ and basically the same $P_{cr}$, and the instability deformation shows the local instability mode above the bearing. This shows that even if the population starting point of the guided genetic algorithm is

**Table 4.3** Comparison of stable optimization results of two different models

| Name of parameter | Model 1 (normal model) | | Model 2 (weak model) | |
|---|---|---|---|---|
| | Initial structure | Optimized structure | Initial structure | Optimized structure |
| $gra\_r_{min}$ | $-5.457 \times 10^{-4}$ | $-2.029 \times 10^{-4}$ | $-6.991 \times 10^{-4}$ | $-2.204 \times 10^{-4}$ |
| $P_{cr}$ (kN/m$^2$) | 3.27 | 5.666 | 3.13 | 5.672 |
| $V$ (m$^3$) | 0.3500 | 0.3724 | 0.3498 | 0.3741 |

different, under the premise of the same other optimization conditions, the guided genetic algorithm can find the optimal solution and converge to the optimal solution, which verifies the robustness of the optimization algorithm.

## 4.3 Collapse Resistance of Model 1 After Stable Optimization

### 4.3.1 Shaking Table Test Process of Model 1 Initial Structure

Model 1 (the normal model) structure exhibits small vibration amplitudes and no obvious deformation when PGA is 100 gal. When PGA reaches 200 gal, the lower part of the structure between C and D platforms first appears with a small number of curved members. The structure is still in the elastic state for the most part, as shown in Fig. 4.17b. As PGA increases from 250 to 350 gal, the second to fifth rings (counted from the bottom up) of the lower part between C and D platforms of the structure gradually bend in an oblique direction, with a clear development and expansion of the range. Materials gradually enter the plastic state, as shown in Fig. 4.17c. When PGA reaches 400 gal, the above-mentioned failure region expands rapidly, and the joint displacement increases significantly. The structure collapses in a short time, as shown in Fig. 4.17d. The maximum displacement of Model 1 as a function of PGA is shown in Fig. 4.18. It can be seen that the PGA-Δ relationship is a double hyperbola. After the load reaches 200 gal, the slope of the curve decreases, and the initial structure shows a certain precursor of collapse failure mode.

Using the display dynamic module LS-DYNA of the general finite element software ANSYS, numerical simulations were carried out on Model 1. Beam161 elements were used to simulate the members, and Mass166 elements were used to simulate the joints. The mass of the joints in the finite element model was the same as the additional mass block of the actual model. The material parameters were taken from the material of the test model, which was Q235 steel, with an elastic modulus of $E = 206 \times 10^3$ N/mm$^2$, a yield strength of $f_y = 235$ N/mm$^2$, a Poisson's ratio of $v = 0.3$, an ideal elasto-plastic model, and an ultimate strain of 0.3. The seismic waveforms of the four tabletops (shown in Fig. 4.4) were also consistent with the test.

## 4.3 Collapse Resistance of Model 1 After Stable Optimization

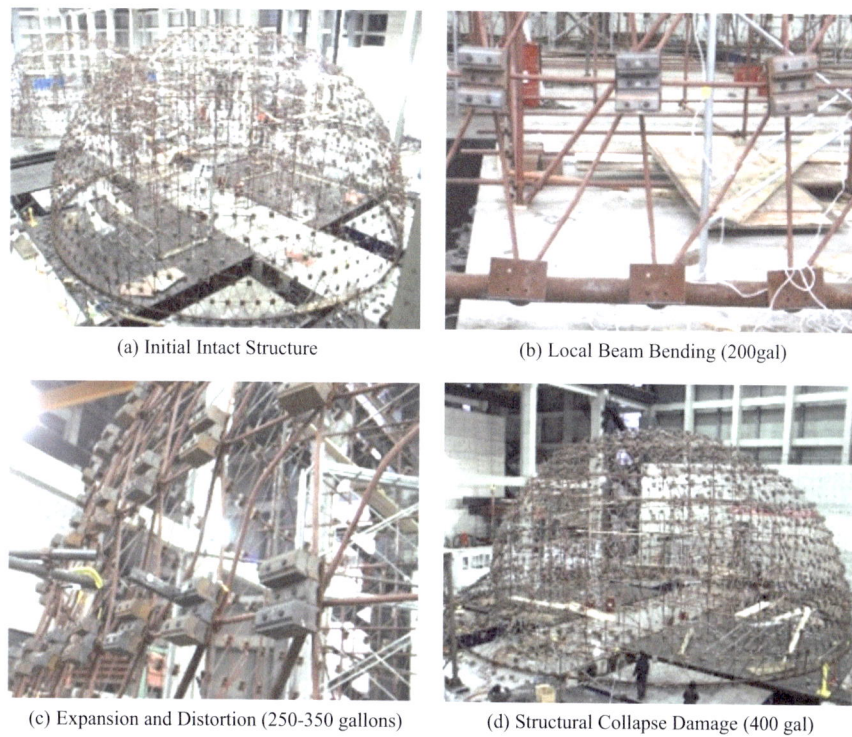

**Fig. 4.17** Model 1 (normal) shaking table test procedure

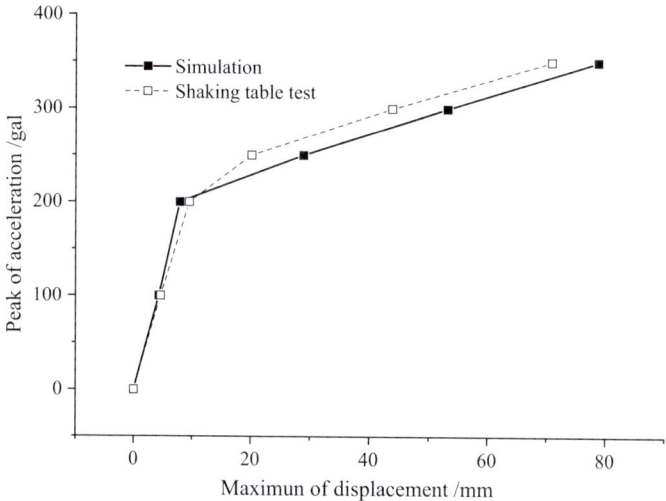

**Fig. 4.18** Model 1 structural load–displacement curve

Through numerical simulation, the relationship between the maximum displacement of the structure and PGA was obtained, as shown in Fig. 4.18. By comparing the PGA-$\Delta$ curves obtained from numerical simulation and experiment, it can be concluded that: (1) Numerically, the simulation results are basically consistent with the experimental results; (2) In terms of curve shape, both show a double-line shape, and the slope of the two curves becomes significantly smaller after PGA reaches 200 gal. By comparing the simulation results with the experimental results, the correctness of the numerical simulation was verified.

### 4.3.2 Collapse Resistance of Model 1 After Stable Optimization

After stabilizing and optimizing Model 1, a multi-point seismic analysis was conducted with varying PGA until the structure collapsed. The seismic performance analysis was conducted using the LS-DYNA display dynamics module in the ANSYS general finite element software. The finite element model and parameter selection were the same as those in Sect. 4.4.1. The seismic waveforms of the four pedestals (shown in Fig. 4.4) were exactly the same as those in the test.

Optimized structure, during the loading process of PGA up to 400 gal, no noticeable bending of members was observed; during the loading process of PGA up to 500 gal, only four supports above, near the main rib of the 2nd–4th ring diagonal brace began to show bending; during the loading process of PGA up to 600 gal, the bending members further developed, and the 2nd–4th ring diagonal braces above four supports had varying degrees of bending; during the loading process of PGA up to 750 gal, the 3rd ring diagonal brace above B and D supports near the main rib bulged out, and the 5th ring diagonal brace was concave, showing a wavy deformation; during the loading process of PGA up to 800 gal, the structure collapsed from the concave area above B and D supports. Figure 4.19 shows the measured load–displacement curve of the initial structure and the simulated load–displacement curve of the optimized structure. It can be seen that the optimized structure collapsed at PGA of 800 gal, while the unoptimized test model 1 collapsed at PGA of 400 gal; from the trend of structural deformation, as the PGA increased, the slope of the PGA-$\Delta$ curve gradually decreased, and before the collapse, the structure deformed significantly, showing a clearly evident collapse failure mode.

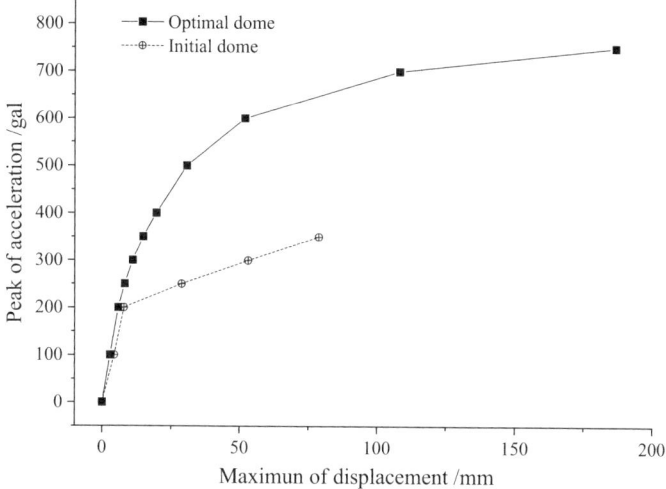

**Fig. 4.19** Model 1 optimized structural loading-displacement curve

## 4.4 Collapse Resistance of Model 2 After Stable Optimization

### 4.4.1 Shaking Table Test Process of Model 2 Initial Structure

Model 2 (the weak model) initially had a stable vibration with no obvious deformation when subjected to a PGA of 100 gal. The structure was in an elastic vibration state. As the loading was increased to 200 gal, the overall vibration of the structure intensified, but local deformation was still not obvious, and no members buckled. When the loading was increased to 250 gal, the pre-designed weak area in the structure suddenly collapsed, while the other parts of the structure still showed no obvious deformation, as shown in Fig. 4.20b. As the loading was increased to 300 gal, the displacement of the collapsed area increased noticeably, and the collapsed area expanded slightly, but the other parts of the structure still showed no obvious deformation, as shown in Fig. 4.20c. When the loading was increased to 350 gal, the structure suddenly collapsed, as shown in Fig. 4.20d. The maximum displacement of Model 2 as a function of PGA is shown in Fig. 4.21. Figure 4.21 shows that when the initial structure was subjected to a PGA of 200 gal, the overall deformation was small. As the PGA was increased further, the deformation of the structure increased dramatically, and the slope of the PGA-$\Delta$ curve suddenly decreased, showing an unpredictable, sudden collapse pattern.

Similarly, in the ANSYS platform, the numerical simulation of model 2 was carried out using the LS-DYNA software, and the relationship between the maximum displacement of the structure and PGA was obtained, as shown in Fig. 4.21. The

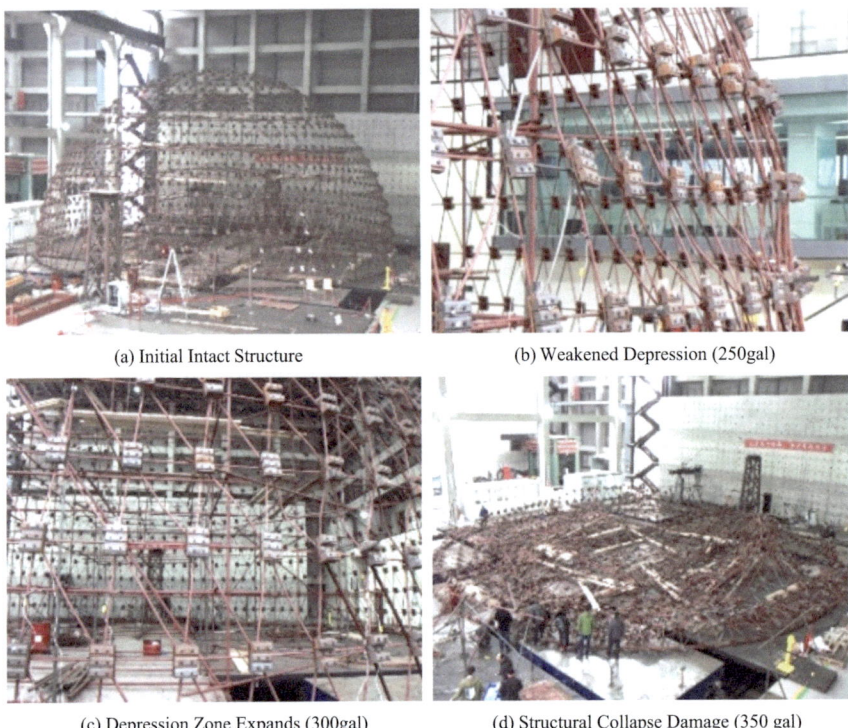

(a) Initial Intact Structure  (b) Weakened Depression (250gal)

(c) Depression Zone Expands (300gal)  (d) Structural Collapse Damage (350 gal)

**Fig. 4.20** Model 2 (weak model) shaking table test procedure

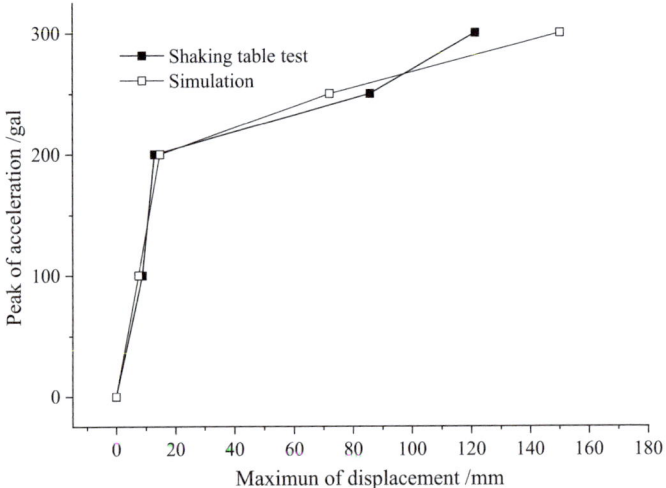

**Fig. 4.21** Model 2 structural load–displacement curve

maximum deformation of the actual structure is basically consistent with the simulated maximum deformation of the structure. At the same time, the shapes of the two curves are basically similar: before 200 gal, the slope of the two PGA-Δ curves is relatively large; after 200 gal, the slope of the two curves suddenly decreases. By comparing the numerical simulation results with the experimental results, the correctness of the numerical simulation was verified.

### 4.4.2 Collapse Resistance of Model 2 After Stable Optimization

Perform multi-point seismic analysis for the optimized model 2 with stable performance, gradually increasing the PGA until the structure collapses. The material and initial structure of model 2 are identical, with the same parameters as those in Sect. 4.4.2. The seismic waveforms of the four tabletops (show in, Fig. 4.4) are also consistent with the test results.

Optimized structure, when loaded to 450 gal, showed no noticeable bending of members, and the structure vibrated smoothly; when loaded to 550 gal, the second ring of diagonal members and the third ring of ring members (counted from the bottom up) above C support began to show signs of bending; when loaded to 650 gal, the bent members above C support continued to develop, and the same position above A support also showed bent members; when loaded to 700 gal, the structure collapsed from above C support. Figure 4.22 shows the measured load–displacement curve of Model 2's initial structure and the simulated load–displacement curve of the optimized structure. It can be seen that the optimized structure's collapse PGA is 700 gal, while the unoptimized experimental model 2's collapse PGA is 350 gal. From the trend of structural deformation, as the peak acceleration increases, the structure's displacement gradually increases, and at the critical collapse point, the structure undergoes significant deformation and shows a predictable collapse pattern.

## 4.5 Chapter Summary

The stability optimization design of two large single-layer gridshells further verified the stability optimization design method proposed. Each gridshell has 3660 members, resulting in a huge number of optimization variables. To fully consider the actual engineering needs, the number of candidate sections (i.e., the range of values for the optimization variables) reached 575; based on the limitations of the experimental conditions, the support constraints were only 40 joints on the supports, and they were unevenly distributed. For the large single-layer gridshell stability optimization problem with huge volume, high design dimension, and unideal support constraint conditions, the optimization design method can provide an optimized solution. Under

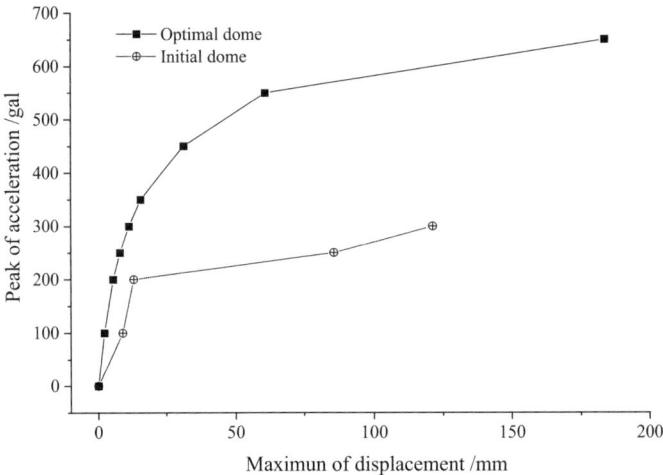

**Fig. 4.22** Structural loading-displacement curves before and after model 2 optimization

the premise of a maximum steel usage limit, the stable bearing capacity of the optimized structure was increased by 1.732 times (model 1) and 1.812 times (model 2). This shows that the stability optimization design method can handle the stability optimization of large and complex gridshells under various support conditions and has good applicability. On the other hand, the two initial large single-layer gridshells have the same support conditions, topological connections, and steel usage, but slightly different member section distributions, and their corresponding two structures have similar steel usage, $gra\_r_{min}$, and $P_{cr}$, and both show local instability modes above the supports. This indicates that, under the same optimization conditions, the optimization design method can find the optimal solution and converge to the optimal solution regardless of the initial structure specified, verifying the robustness of the optimization algorithm.

The stably optimized structure not only has good static stability, but also has good collapse resistance. The PGA corresponding to the collapse of the stabilized optimized structure is 2 times higher than that of the original structure. In terms of collapse mode, the load–displacement curve of the initial structure of model 1 is double broken line, showing a collapse mode with certain symptoms; the load–displacement curve of the initial structure of weak model 2 has an obvious inflection point, showing a collapse mode with no symptoms and sudden failure; and the structural displacement of the two corresponding optimized models gradually increases with the increase of the peak acceleration. At critical collapse, the structure deforms significantly and presents a collapse mode with obvious signs. Therefore, both the critical load and the collapse mode of the optimized structure are obviously better than that of the initial structure.

# References

1. Lu H (2015) Experimental study on collapse of single-layer spherical shell with multi-point input. Southeast University
2. Ye J, Li K (2016) Redundancy characteristics of single-layer spherical lattice shell based on response sensitivity under multi-point input. Chin J Civil Eng 49(9):20–29
3. National Standard of the People's Republic of China (2008) GB/T 17395–2008 dimension, shape, weight and allowable deviation of seamless steel pipes. Standards Press of China, Beijing

**Open Access** This chapter is licensed under the terms of the Creative Commons Attribution-NonCommercial-NoDerivatives 4.0 International License (http://creativecommons.org/licenses/by-nc-nd/4.0/), which permits any noncommercial use, sharing, distribution and reproduction in any medium or format, as long as you give appropriate credit to the original author(s) and the source, provide a link to the Creative Commons license and indicate if you modified the licensed material. You do not have permission under this license to share adapted material derived from this chapter or parts of it.

The images or other third party material in this chapter are included in the chapter's Creative Commons license, unless indicated otherwise in a credit line to the material. If material is not included in the chapter's Creative Commons license and your intended use is not permitted by statutory regulation or exceeds the permitted use, you will need to obtain permission directly from the copyright holder.

# Chapter 5
# Topology Optimization Design of Joints of Single-Layer Gridshells

**Abstract** A single-layer gridshell is a shape-resistant structure, where the structure's shape and grid division are designed in a way that reduces the bending effect caused by loads, allowing members to bear axial forces as much as possible to fully utilize the material's potential. This results in a lightweight and transparent structure that spans large distances. As the intersection points of members, joints not only have complex loading conditions but also have a significant impact on the structure's overall performance. First, the joint's stiffness affects the distribution of force flow, thereby affecting the structure's stiffness, static stability, and other mechanical properties. Second, if a joint fails, force flow will be interrupted, causing a change in the transmission path, and may even lead to the collapse of the entire structure. Therefore, the current design code recommends only a limited number of joint types, and joint size design tends to be conservative. In summary, there is an optimization space for the joints of single-layer gridshells under the premise of ensuring their safety and reliability. This chapter begins by reviewing the basic components of a joint, clarifying the core part of the joint optimization design—the joint core. In order to improve the rotational stiffness of the joint, a rotational stiffness optimization model of the joint core was established, and the topology optimization method was used to optimize the joint core under the premise of meeting all design requirements. At the same time, a universal connection interface was designed by combining structural design, which can meet the connection requirements of the members in different orientations. In order to improve the safety performance of the joint, a safety performance evaluation index of the joint was first defined. Based on this index, a joint safety performance optimization model was established, and an optimization algorithm program was compiled to perform topology optimization on two-dimensional joints.

## 5.1 Basic Components of Space Structure Joints

Understanding the basic components of spatial structures is the foundation for optimizing joint design. Liu [1, 2], Yin and Liu [3], Hanaor [4], Weng [5], and others have reviewed joints in spatial structures from different angles. Arciszewski and Uduma [6] have conducted a thorough study of the joints in space steel structures. When the joint is connected in the most complex form, it can be decomposed into eight basic components, as shown in Fig. 5.1. The internal forces of the members are transmitted through connection ⑦ to the joint end ②, and finally, through connection ⑥, to the joint core formed by elements ① and ⑤. Ultimately, all the member end forces converge at the joint core, forming a balanced force system.

Because all the end moments converge at the joint core, the joint core is subjected to complex loading; at the same time, the main deformation of the joint also occurs at the joint core, so the joint core has the greatest influence on the joint stiffness. Under the premise that all connections are safe and reliable, the mechanical properties of the joint core are crucial to the overall performance of the joint. The joint end (element ②) is connected to the joint core through connection ⑥ on one end and to the end of the beam element through connection ⑦ on the other end. Therefore, the form of the joint end determines the specific connection method, the method of joint installation, and the construction technology of the overall structure. The rational selection of the

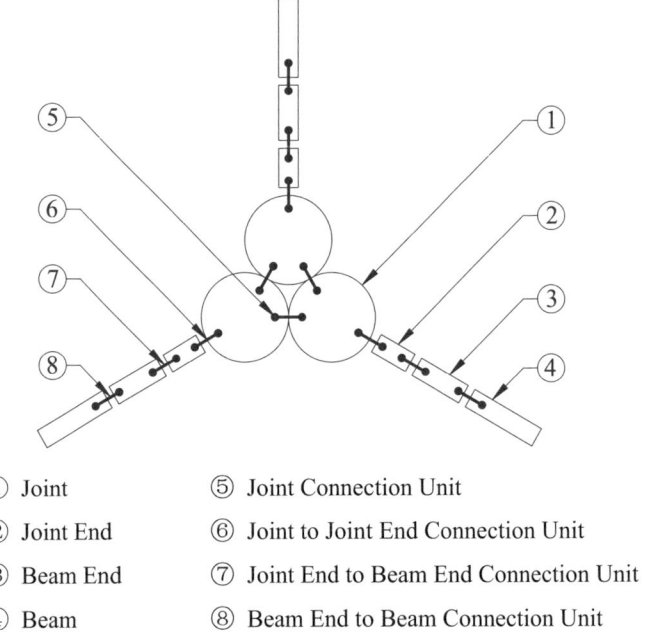

① Joint  　　　　　　⑤ Joint Connection Unit
② Joint End  　　　　⑥ Joint to Joint End Connection Unit
③ Beam End  　　　　⑦ Joint End to Beam End Connection Unit
④ Beam  　　　　　　⑧ Beam End to Beam Connection Unit

**Fig. 5.1** Basic components of a joint system

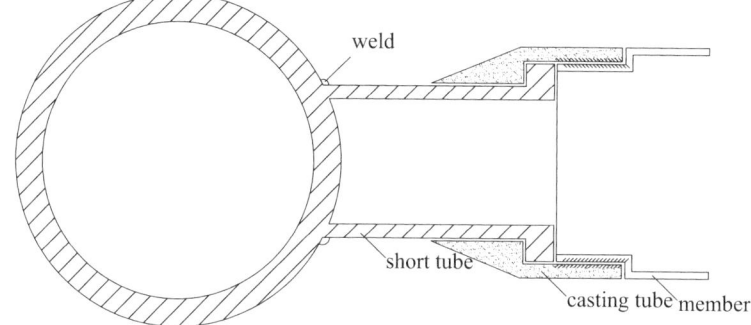

**Fig. 5.2** Okta-S joint

joint end not only requires consideration of the mechanical properties, but also the construction technology, and should be combined with engineering experience.

For example, explain in detail the basic components of the Okta-S joint. The Okta-Platte system was developed by the German Mannesmann company in 1959 [7]. However, the field welding workload for this joint is large. In order to reduce the field welding workload, the company developed the Okta-S joint with a sleeve in 1974, as shown in Fig. 5.2.

In the Okta-S joint, a short pipe is pre-welded to the ball joint, connected to the member through a threaded sleeve (element 3) and a rod (element 4), which greatly reduces the welding workload. The joint (element 1) is a hemispherical shell that is welded (element 5) to form the spherical joint nucleus. If the diameter is large, reinforcing ribbed diaphragms (element 5) are set. The joint end (element 2) is a short pipe that is welded (element 6) to the joint nucleus. The threaded rod end (element 3) is prefabricated with the member (element 4). The sleeve has a thread on the inner wall of one side of the sleeve, which is engaged with the thread of the rod end. On the construction site, the threaded rod end is connected to the short pipe through the sleeve (element 7) to form a complete joint.

## 5.2 Topology Optimization of Space Structure Joints Based on Rotational Stiffness

### 5.2.1 Joint Construction Design

Space structure expert Makowski [8] pointed out that the development of joints should be combined with the given structural form and be designed specifically. Makowski illustrated this with an example, showing that "universal joints" suitable for various structures are often complex and expensive to construct. On the other hand, even if joints are developed specifically for a certain type of structure, if they are not well designed, they will gradually be phased out in practice, such as the Nodus joint in the UK [9]. Therefore, for a single-layer shell structure with circular tube section members, joint detail design should be carried out to meet the actual engineering needs.

Document [5] combines the functional and market demands of joints to propose five basic principles for new joints, namely structural suitability, geometric adaptability, tolerance for errors, aesthetic effects, and comprehensive economic efficiency. Among them, structural suitability refers to functional requirements such as bearing capacity and stiffness of joints. Specifically, the geometric adaptability and tolerance for errors involve design requirements. Geometric adaptability means that as the single-layer grid structure's shape becomes increasingly diverse, the connection between spatial members becomes more complex, and the joint should have good adaptability to the cutting angle projection, cutting angle inclination, and axis rotation angle of the connected members. The geometric adaptability of the joint determines its market potential. Tolerance for errors means that the joint can effectively control and adjust for deviations that occur during manufacturing and installation. The tolerance for errors of the joint is related to the safety of the entire structure and the construction schedule.

For the joint and circular tube members discussed in this book, the following structural requirements are supposed to be considered.

(1) Directionality-free requirements: The mechanical properties of the joint will not change due to the different connection orientations of the members. Directionality-free joints are easy to construct, eliminating the possibility of changes in mechanical properties due to incorrect high-altitude construction, and are safe and reliable. Typical directionality-free joints include the Mero standard joint and the welded hollow ball joint.
(2) Geometric adaptability: No matter how complex the connection interface, cutting plane projection angle, cutting plane inclined foot, and axial rotation angle of the member cause the spatial position of the member to be, the joint should be able to connect with the member tightly and reliably.
(3) Assisted centering requirements: The joint should have relevant measures to ensure that the axis of the connected members intersect at the joint's centroid. Because the assembly of space structures is sometimes carried out at high altitudes, aligning the members is difficult. Member misalignment will cause secondary stresses at the joint and create structural defects, significantly reducing the stability of the single-layer dome structure. Therefore, the joint should have a construction that assists in aligning the members.
(4) Member adaptability: A joint should be able to reliably connect with multiple members of similar cross-section specifications. First, joints in space structures are generally connected to multiple members, and when these members have different cross-section dimensions, the joint should be able to adapt to all connected members. Secondly, in a grid structure, reducing the number of joint types within a certain range can make construction easier, achieve mass production, and lower costs.

Unlike trusses, spatial trusses, and double-layer shells, in single-layer shell structures, all the members intersecting at a point are approximately in one plane, as shown in Fig. 5.3. Therefore, a connection interface perpendicular to the member plane is not necessary. To meet the connection requirements of circular tube members, a initial joint design as shown in Fig. 5.3 is used. The green part is the joint kernel to be optimized, and the yellow part is the joint end. The joint end is a partial spherical shell centered at the joint center of mass. Because the joint end is a spherical surface in all directions, it can connect members coming from different directions, even in cases where the spatial distribution of the members is irregular (see Fig. 5.4). The spherical joint end can also achieve convenient and precise connections in such cases. In the processing of circular tube members, as long as the end section of the member is perpendicular to the member axis, the spherical connection interface and the circular tube member can be tightly connected, and the member axis must intersect at the center of the sphere, avoiding alignment errors. This not only avoids the secondary stress of the joint, but also can greatly reduce the impact of defects on the shell structure. In theory, as long as the outer diameter of the circular tube member is not greater than the height of the joint end, it can be connected to the joint for different sizes of circular tube members, which makes it possible for: (1) When the outer diameters of the members intersecting at the same joint are different, the joint determined by the maximum outer diameter of the members can effectively connect all the connected members. (2) In the entire dome structure, the number of joint types can be reduced within a certain range, thereby reducing production costs and making construction easier. In summary, by constructing the spherical joint end, a convenient, reliable, and precise connection has been achieved, meeting the requirements of the applicability and geometric adaptability of the members.

**Fig. 5.3** Initial joints

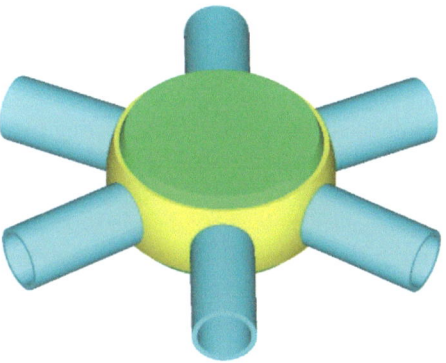

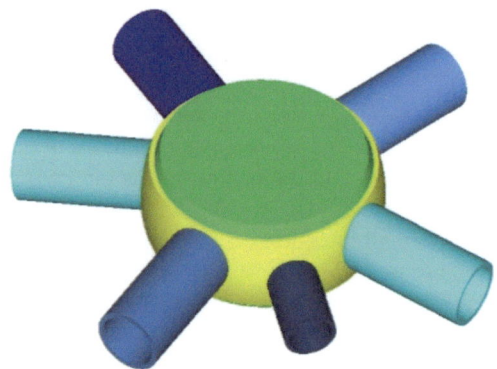

**Fig. 5.4** Connections between joints and members with different sizes and orientations

## 5.2.2 Topology Optimization Model for Joint Nucleus Rotational Stiffness

Professor Luo and Academician Shen [10] studied the influence of joint elements on the bearing capacity of mesh structures and pointed out that the size of the joint element can be ignored when it is less than 5% of the rod length. Liu et al. [11] found that when the diameter of the welded ball joint is less than 12% of the geometric length of the member, the joint has a small effect on the axial stiffness of the structure, but a large effect on the bending stiffness. Generally, new types of joint structures have novel and efficient designs, and the size of the joint can be ignored for the overall performance of the structure. Cao et al. [12] pointed out that the torsional stiffness of the joint has only a 0.4% impact on the ultimate load of the single-layer mesh structure and can be ignored for the stability performance. Therefore, the relevant research is mainly focused on the influence of joint bending stiffness on the stability performance of mesh structures. Fan et al. [13] reviewed historical literature and conducted extensive case analysis, concluding that joint bending stiffness significantly affects the stable bearing capacity of single-layer mesh structures.

Traditional frame structures have distinct principal and secondary axes, so the nodal stiffness can be decomposed into rotational stiffness about the strong axis and rotational stiffness about the weak axis. It is difficult to represent the nodal stiffness of a single-layer shell structure using orthogonal rectangular coordinates, and it can be represented by the normal vector and tangent plane of the surface at the joint. The rotational stiffness of the joint in the tangent plane is defined as the in-plane stiffness of the shell, and the stiffness perpendicular to this direction is called the out-of-plane stiffness of the shell. Research on the weak-axis stiffness of steel structure joints has been less studied, for example, Nguyen and Kim [14] could only assume that the weak-axis stiffness of the joint is one-fifth of the strong-axis stiffness for research. In current research on spatial structure joints [15–17], experimental and numerical simulations are all focused on the out-of-plane bending performance of the joint shell, and the in-plane bending performance of the joint shell is still very limited [18]. For

## 5.2 Topology Optimization of Space Structure Joints Based on Rotational ...

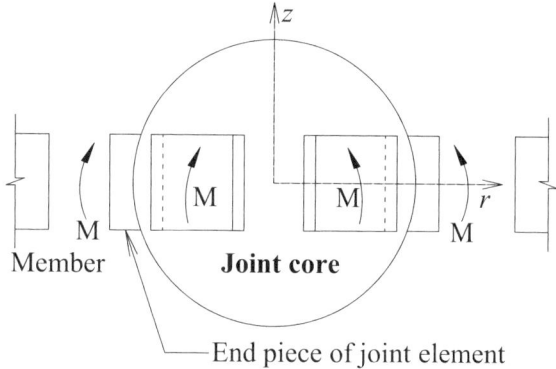

**Fig. 5.5** The out-of-plane bending moment at the joint $M_y$

spherical shell structures, the out-of-plane stiffness of the joint is dominant [13]; for cylindrical shell structures, the in-plane stiffness of the joint is dominant [19]; for single-layer free-form structures, there is no general rule. Therefore, the joint in a single-layer grid structure should have sufficient in-plane rotational stiffness and out-of-plane rotational stiffness to transmit moment in any direction.

To express the out-of-plane stiffness of a joint, a unit out-of-plane moment is applied at each joint end, as shown in Fig. 5.5, forming a self-balanced force system $M_y$. The strain energy of the joint area is denoted as $C_{My}$. $C_{My}$ represents the flexibility of the joint in the force system $M_y$, which is the reciprocal of the stiffness. To express the in-plane stiffness of a joint, a unit in-plane moment is applied at each joint end, and the moment directions of adjacent joint ends are opposite, as shown in Fig. 5.6, forming a self-balanced force system $Mz$. The strain energy of the joint area is denoted as $C_{Mz}$.

To improve the rotational stiffness of the joints, it is equivalent to reducing the flexibility of the joint area (equivalent to reducing the strain energy of the joint). When using the variable density method [20] for topology optimization, the stiffness optimization model of the rigid joint in the single-layer shell structure can be expressed as follows:

$$
\begin{aligned}
&\text{Find}: \boldsymbol{\rho} = [\rho_1, \rho_2 \ldots \rho_i \ldots \rho_n]^T \in R^n \\
&\text{Min}: 1 \times C_{My}(\boldsymbol{\rho}) + 1 \times C_{Mz}(\boldsymbol{\rho}) \\
&\text{s.t.}\ \mathbf{F} = \mathbf{K}(\boldsymbol{\rho})\mathbf{U} \\
&\quad m(\boldsymbol{\rho}) \leq m_{\text{thres}} \\
&\quad \sigma_{\max}(\boldsymbol{\rho}) \leq f \\
&\quad 0 \leq \rho_i \leq 1
\end{aligned}
\tag{5.1}
$$

In the equation: the optimized variable $\boldsymbol{\rho} = [\rho_1, \rho_2 \ldots \rho_i \ldots \rho_n]^T$ is the vector of relative densities for all elements; $\rho_i$ is the relative density of element $i$; $C_{My}$ and $C_{Mz}$ respectively represent the strain energy of the joint domain under the action of force system $M_y$ and $M_z$; $\mathbf{F}$, $\mathbf{K}$, and $\mathbf{U}$ respectively represent the joint's load vector,

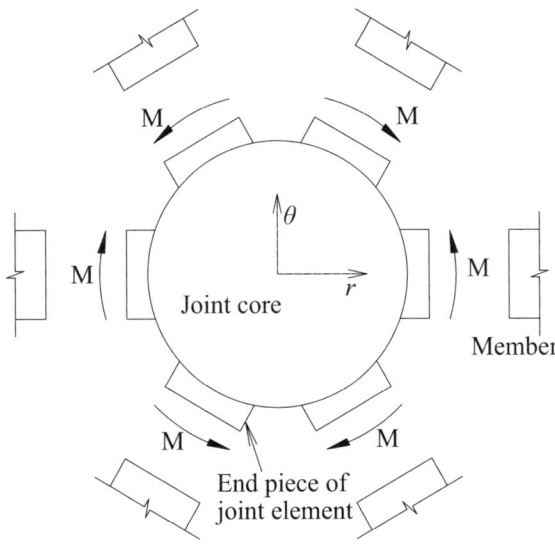

**Fig. 5.6** Moment of bending in the joint plane, $Mz$

the global stiffness matrix, and the displacement vector of the joint; $m(\rho)$ represents the joint mass with respect to the variable $\rho$; $m_{thres}$ represents the upper limit of the joint mass; $\sigma_{max}(\rho)$ represents the maximum equivalent stress of the joint about the variable $\rho$ in the design load condition; $f$ is the material strength design value, and in this paper, Q345 steel is selected as the design material, with the yield strength design value $f = 310$ N/mm$^2$.

### 5.2.3 Key Technologies for Joint Nucleus Topology Optimization (Application of Equivalent Joint Loads)

The results of topology optimization are usually closely related to the force boundary conditions and displacement boundary conditions. A small change in the boundary conditions can lead to completely different optimization results. Therefore, when establishing the analysis model, the various boundary conditions should be correctly reflected to reduce the sensitivity of the optimization results to the boundary conditions and thus obtain stable and reliable topology optimization results.

In the design load condition, when performing a global analysis of truss structures, the face load within the load-bearing area of the joints must be equivalent to a concentrated load $P$ applied at the joint. Taking the joint as the isolated body, the joint isolated body is in equilibrium under the reaction force at the end of the member and the equivalent concentrated load $P$ at the joint, as shown in Fig. 5.7. If the concentrated load $P$ is directly applied at the joint core, for the stress-optimized model of Eq. (5.1), it is equivalent to assuming that there must be material distribution near the loading point in advance, limiting the possibility of exploring a better joint form in topology

## 5.2 Topology Optimization of Space Structure Joints Based on Rotational …

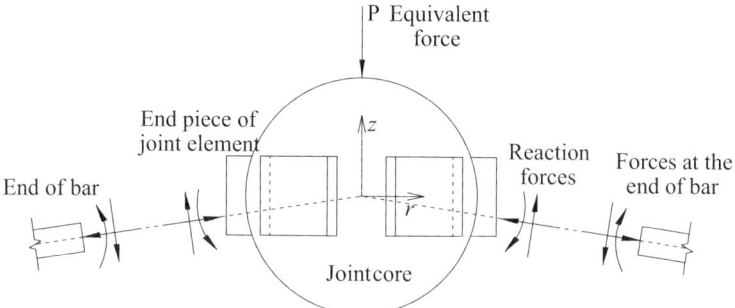

**Fig. 5.7** Equilibrium of joint under equivalent concentrated load and end moment of rod

optimization, and may also lead to stress singularity at the loading point [21], causing stress constraint distortion. If the concentrated load $P$ is not directly applied in the form of force, but is realized through the support reaction force at the constraint point to achieve joint equilibrium, the optimization result depends on the position of the constraint. When the constraint is applied at different positions, the point of action of the support reaction force is different, and the optimization result is different, showing significant boundary sensitivity.

To avoid applying concentrated loads in the design domain and optimizing the results' dependence on boundary conditions, this paper proposes applying an acceleration field in the design space, which is equivalent to a concentrated load through the volume inertia force. Considering that the joint mass $m$ tends to converge to the mass threshold $m_{thres}$ as the optimization tends to converge, the acceleration numerical value $a$ can be calculated according to Eq. (5.2), with the direction opposite to $P$. By applying an inertial force equivalent to a concentrated load with the property of body force, not only can the problem of directly applying concentrated loads in the design domain be avoided, but also the joints can achieve self-balance. The displacement boundary condition is only used to limit the rigid body displacement, reducing the sensitivity of the optimization results to the boundary conditions. The actual optimization results also show that, after processing the concentrated forces at the joint using this method, the optimization results are the same regardless of which end of the joint the constraint is applied.

$$a = \frac{P}{m_{thres}} \tag{5.2}$$

## 5.2.4 Joint-Core Topology Optimization Algorithm

Topology optimization is a scientific method that seeks the optimal structural performance of a limited amount of material distributed in space [22]. For complex structures under load, the results of topology optimization are generally better than those of optimized design based on experience. Among various optimization methods, the SIMP method in the variable density method is the most widely used. In the variable density method, the relative density ρ of the element is used to represent the state of the element. When ρ = 1, it indicates that the element exists; when ρ = 0, it indicates that the element does not exist; at the same time, it allows the relative density of the element to take continuous values between 0 and 1, that is, to allow the existence of intermediate density elements. To reflect the influence of intermediate density elements on the structure, the SIMP method establishes the relationship between the elastic modulus of the element and its density, as shown in Eq. (5.3).

$$E(\rho_i) = \rho_i^p E_0 \tag{5.3}$$

In the formula: $\rho_i$ represents the relative density of unit $i$; $E(\rho_i)$ is the elastic modulus of unit $i$; $E_0$ is the elastic modulus of the solid material; and $p$ is the penalty coefficient, which is usually set to 3.

After establishing the relationship between unit relative density and modulus of elasticity, the optimized model given by Eq. (5.1) can be solved by various methods such as optimality criteria, sequential linear programming, and method of moving asymptotes.

## 5.2.5 Comparison of Joints (Mechanical Properties of Welded Hollow Ball Joints)

The design object is the vertex of the single-layer grid structure with a span of 22 m and an elevation of 11 m selected from Sect. 3.3 of this article. The structure is shown in Fig. 3.2. Under the design load of 3.5 kN/m², after the stability optimization design, the 6 connected rods are all Φ140 × 3.2 round steel tubes, and the length of each rod is 2.463 m. The stress constraint condition of the joint under the design load is taken as the force situation of the structure vertex, and the basic geometric dimensions of the optimized joint are determined by the size of the connected rods.

For single-layer grid structures, there are not many choices of rigid joint forms. In this study, the most common welded hollow ball joint is chosen. According to the design requirements of the Technical Code for Space Grid Structures (JGJ 7-2010) [23], the ratio of the outer diameter of the hollow ball to the outer diameter of the main steel pipe should be 2.4–3.5, so the range of the outer diameter of the hollow ball D is 336–490 mm. To make the comparison more convincing, a smaller joint size is chosen, with $D = 360$ mm. The ratio of the outer diameter of the hollow

**Fig. 5.8** Welded hollow ball joints

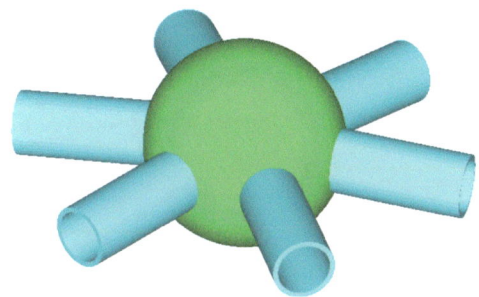

ball to its wall thickness should be 20–35, so the range of the wall thickness t is 10.2–18.0 mm, with $t = 10$ mm. After calculation, the outer diameter of the hollow ball satisfies the requirement of JGJ 7-2010 (Sect. 5.2.6) [23], that is, the minimum clearance between members. Finally, the outer diameter of the welded hollow ball is $D = 360$ mm, the wall thickness is $t = 10$ mm, as shown in Fig. 5.8. The joint mass is 30.22 kg; the strain energy of the welded hollow ball joint under the force system $My$ is $C_{My} = 4.4961 \times 10^{-7}$ J; the strain energy of the joint under the force system $Mz$ is $C_{Mz} = 5.2378 \times 10^{-7}$ J; after calculation, the joint meets the load-bearing requirements.

### 5.2.6 Determining the Optimal Topology for Maximum Stiffness in a Fixed-Quality Structure

Given a maximum mass limit for the joints, the joints that are optimized to achieve the maximum stiffness subject to this constraint are referred to as the fixed-mass stiffness-maximizing topology optimization joints. The mathematical expression of the optimization model is given in Eq. (5.1). The three-dimensional schematic of the initial optimization joint is shown in Fig. 5.3 and the detailed dimensions are shown in Fig. 5.9. The joint core is the topology optimization design domain, and the joint ends are not involved in the topology optimization. The joint end mass is 12.3 kg. Compared with the hollow sphere joint used in Sect. 5.2.5, the maximum geometric size of the initial joint is equal to the hollow sphere joint, while the topology optimization design domain is within the hollow sphere joint, without extending beyond its spatial range.

With the help of Altair Optistruct platform, the joint core mass threshold was set to 12.5 kg, and the topology optimization was performed on the joint core, resulting in the optimized joint shown in Fig. 5.10. In Fig. 5.10, the yellow transparent part represents the constructed joint end. The total mass of the entire joint (including the joint end and the optimized joint core) is 24.8 kg. The bending strain energy of the optimized joint in the plane of the shell and the plane perpendicular to the shell are $C_{My} = 3.5078 \times 10^{-7}$ J and $C_{Mz} = 3.0098 \times 10^{-7}$ J, respectively. Under the

**Fig. 5.9** Initial joint size (Unit: mm)

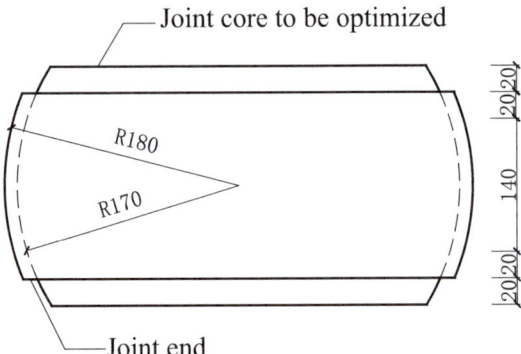

design load, $\sigma_{max} = 159.7 \text{ N/mm}^2 < f$, which satisfies the strength design constraint condition.

Optimizing the joint to rotate symmetrically around the central axis is a typical directionless joint. This indicates that the joint can effectively constrain the rotation of the member in any direction, thus avoiding the need to adjust the spatial orientation

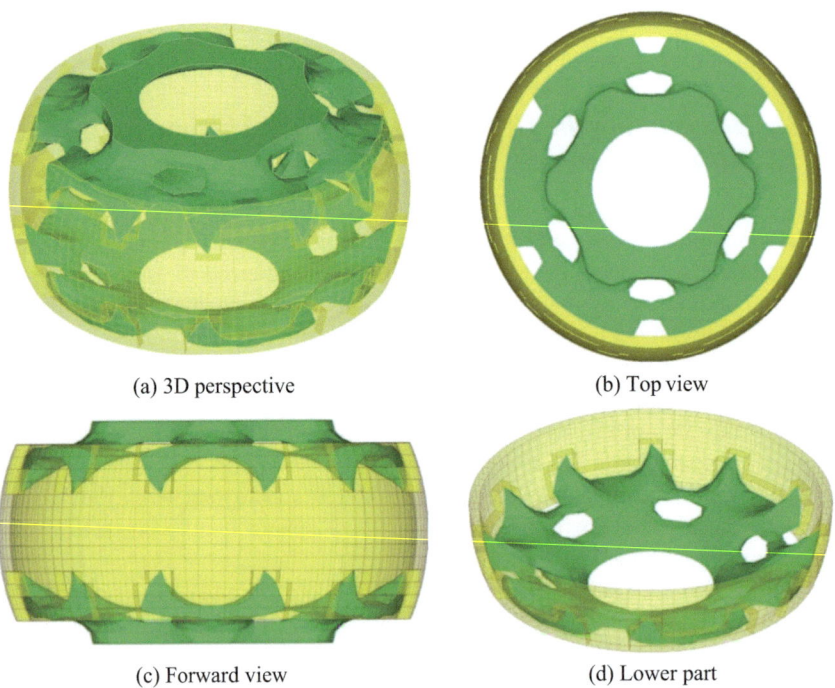

(a) 3D perspective  (b) Top view

(c) Forward view  (d) Lower part

**Fig. 5.10** Determining the optimal topology for maximum stiffness in a fixed-quality structure

of the joint during construction. Under the condition of meeting the strength requirement, the optimized joint provides sufficient rotational stiffness in two directions. Compared with the welded hollow ball joint, the total mass of the optimized joint is 24.8 kg, which is less than the 30.22 kg of the hollow ball joint; in both directions, the strain energy ($C_{My} = 3.5078 \times 10^{-7}$ J, $C_{Mz} = 3.0098 \times 10^{-7}$ J) of the optimized joint is less than the strain energy of the welded hollow ball joint ($C_{My} = 4.4961 \times 10^{-7}$ J, $C_{Mz} = 5.2378 \times 10^{-7}$ J), indicating that the stiffness of the optimized joint is greater than that of the welded hollow ball joint.

In the maximum stiffness topology optimization model with a fixed mass, the optimization goal is to minimize the sum of $C_{My}$ and $C_{Mz}$. Although the weights of $C_{My}$ and $C_{Mz}$ in Eq. (5.1) are the same, it is impossible to precisely control the in-plane strain energy and out-of-plane strain energy during the optimization process, resulting in significant differences in the final optimized joint's $C_{My}$ and $C_{Mz}$.

### 5.2.7 Optimal Topology Design for Minimum Stiffness and Mass of Fixed Joints

In optimization design, more often than not, we know a certain type of performance requirement for the component to be optimized (such as stiffness, frequency, minimum characteristic value buckling, etc.), and the goal is to optimize the design with the lowest possible weight. If we know the requirement for the joint stiffness of the entire structure, we can perform a fixed-stiffness minimum-mass topology optimization, as shown in Eq. (5.4).

$$\begin{aligned}
&\text{Find}: \boldsymbol{\rho} = [\rho_1, \rho_2 \ldots \rho_i \ldots \rho_n]^T \in R^n \\
&\text{Min}: m(\boldsymbol{\rho}) \\
&s.t.\ \mathbf{F} = \mathbf{K}(\rho)\mathbf{U} \\
&\quad C_{My}(\boldsymbol{\rho}) \leq [C_{My}] \\
&\quad C_{Mz}(\boldsymbol{\rho}) \leq [C_{Mz}] \\
&\quad \sigma_{\max}(\boldsymbol{\rho}) \leq f \\
&\quad 0 \leq \rho_i \leq 1
\end{aligned} \quad (5.4)$$

In the equation, $[C_{My}]$ and $[C_{Mz}]$ are the upper bounds of strain energy in the plane outside and inside the shell, respectively. The other symbols are the same as in Eq. (5.1).

Setting the strain energy upper limit is equivalent to setting the minimum stiffness. In the optimization model of Eq. (5.4), all the performance indicators of the joints are clearly controlled. When the structure has different demands for the in-plane stiffness and out-of-plane stiffness of the joints, it can also achieve the topology optimization design of joints with different stiffness requirements.

**Fig. 5.11** Initial joint size after deleting the nucleus of the intermediate joint (Unit: mm)

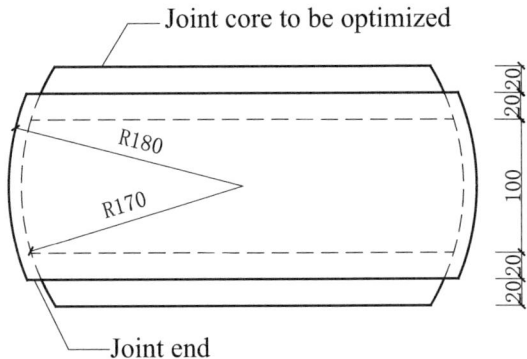

From the relevant computational experience and structural concept analysis in Sect. 5.2.6, it can be concluded that the optimized joint core is a symmetrical structure above and below, and there is usually no material distribution in the middle of the joint. Considering that the efficiency of topology optimization drops sharply as the size of the finite element model increases, some intermediate elements were deleted in the design domain of this section to improve optimization efficiency. The geometric dimensions of the initial structure are shown in Fig. 5.11.

Using the Altair Optistruct platform, a topology optimization design with minimum stiffness and mass for the initial joints shown in Fig. 5.11 was conducted. The upper limits of joint strain energy $[C_{My}] = 2.5 \times 10^{-7}$ J, $[C_{Mz}] = 2.5 \times 10^{-7}$ J were used, resulting in the optimized joints shown in Fig. 5.12. The core mass of the joint is 14.0 kg, and the total weight (including the joint end and the joint core) is 26.8 kg. After optimization, $C_{My} = 2.4976 \times 10^{-7}$ J $\leq [C_{My}]$, $C_{Mz} = 2.2959 \times 10^{-7}$ J $\leq [C_{Mz}]$, which satisfies the stiffness constraint conditions. Under the design load condition, $\sigma_{max} = 240.5$ N/mm$^2$ < f, which satisfies the strength design constraint conditions. Figure 5.12a shows that, without human intervention, the topology optimization-obtained joint core transitions smoothly and continuously with the joint end, and the force transmitted by the beam to the joint end can be transmitted to the joint core, avoiding stress concentration problems.

Under the influence of the out-of-plane moment $M_y$, the stress distribution at the optimized joint is shown in Fig. 5.13. Under the action of $M_y$, the upper half of the joint core is subjected to compression and the lower half is subjected to tension. Both the upper and lower halves of the joint are subjected to forces, which are transmitted from the radial elements to the central ring-shaped hub structure. Similar to arch structures, the ring-shaped hub structure is almost uniformly loaded throughout its cross section, balancing the load from the six joint ends. Figure 5.13 shows that the stress distribution of the central ring-shaped hub structure is basically uniform, verifying the conceptual analysis results of the structure's load-bearing capacity.

Under the action of the moment $M_z$ in the plane, the stress distribution in the upper and lower parts of the joint is optimized to be the same, and the distribution of the joint core stress and the flow of force can be seen in Fig. 5.14. Under the action of

## 5.2 Topology Optimization of Space Structure Joints Based on Rotational …

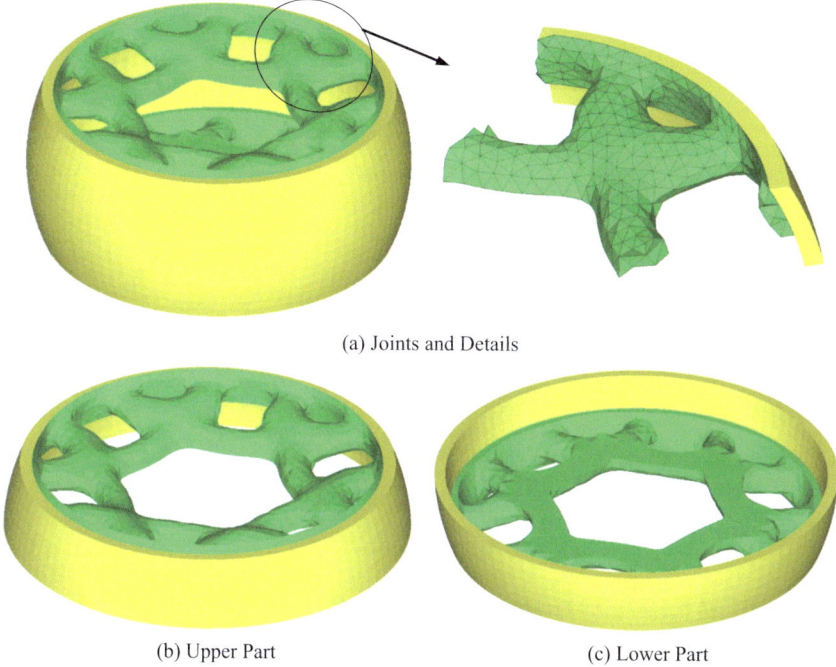

(a) Joints and Details

(b) Upper Part

(c) Lower Part

**Fig. 5.12** Optimal topology with minimum stiffness and maximum mass

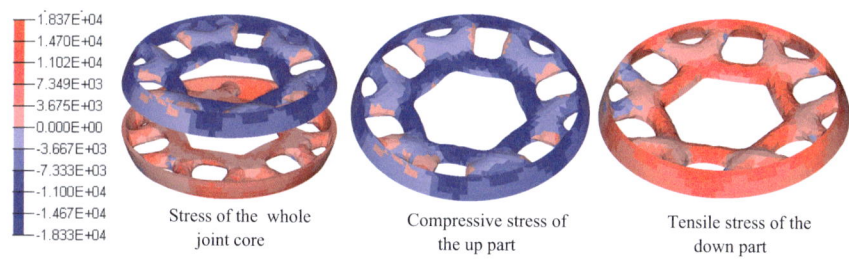

Stress of the whole joint core

Compressive stress of the up part

Tensile stress of the down part

**Fig. 5.13** Optimizing stress distribution in $M_y$ subject joint

the unit moment in the plane, one side of the joint end is subjected to tension and the other side is subjected to compression. The flow of force on the tension side of the joint end passes through the arched connection and balances the tension force with the adjacent joint end, as shown by the red arrow in Fig. 5.14b; similarly, the pressure on the compression side of the joint end also passes through the arched connection and balances the pressure with the adjacent joint end, as shown by the blue arrow in Fig. 5.14b. At the same time, these tension arched connections intersect with each other while supporting each other to maintain the uninterrupted flow of force and improve the structural stiffness.

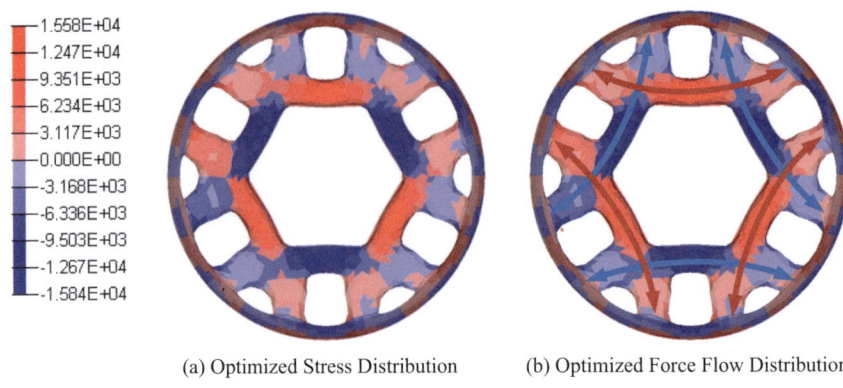

(a) Optimized Stress Distribution  (b) Optimized Force Flow Distribution

**Fig. 5.14** Optimizing stress distribution and force flow distribution under $Mz$ condition

## 5.3 Topology Optimization Based on Safety Performance of Joints

### 5.3.1 Joint Safety Performance Indicators

The criteria for evaluating the structural safety are generally divided into two categories [24]. The first category is stochastic indicators, which are risk evaluations based on probability considerations. The second category is deterministic indicators, which are usually determined based on the structure's response under a given load to determine the structural safety index. Literature [25] and [26] provide a detailed list of various structural safety evaluation indicators.

For many random indicators, there is no universal and widely accepted standard indicator. At the same time, for some extreme actions, such as the impact of 911 and the explosion of the federal building in Oklahoma City, it is difficult to establish probability models for related actions. Therefore, there are certain limitations in the random safety evaluation index. In the second method, the safety assessment results are often dependent on the loading conditions. A structurally safe structure under one loading condition may not be safe under another loading condition. For single-layer gridshells, consideration must be given to wind loads, seismic actions, and semi-span distributed snow loads in different directions, which leads to a large number of loading conditions in the analysis of gridshells and joints. The safety assessment method based on load response not only has a large workload, but is more likely to lead to unreconcilable safety assessment results.

In deterministic indicators, there is a class of indicators that are independent of the external load and only related to the structure itself. This independent load-free deterministic safety assessment method focuses on the structural topology connection, stiffness distribution, etc., which are intrinsic characteristics of the structure,

## 5.3 Topology Optimization Based on Safety Performance of Joints

and emphasizes the inherent robustness of the structure. Typical independent load-free safety assessment methods include the configuration fragility method described in Chap. 2 of this paper [27, 28], the method proposed by Starossek and Haberland based on the eigenvalue of the stiffness matrix [26] and so on.

Whether the structure degenerates from a rigid structure to a mechanism due to broken members, loses stability due to compression, collapses due to failed supports, forms plastic hinges due to material entering the plastic state, or experiences other forms of failure, at the critical point of structural failure, the stiffness matrix of the structure will change from nonsingular to singular. If the initial stiffness matrix of the structure is far away from the singularity boundary, then the structure itself has good safety performance. Demmel [29] found that the shortest distance from the singularity boundary of the matrix $\mathbf{K}$ is inversely proportional to the state function $\kappa(\mathbf{K})$. Therefore, the safety evaluation index of the structure $\delta_s$ is defined as follows:

$$\delta_s = \frac{1}{\kappa(\mathbf{K})} \tag{5.5}$$

In the formula: $\kappa(\mathbf{K})$ is the state function of matrix $\mathbf{K}$, which is calculated according to the following formula:

$$\kappa(\mathbf{K}) = \frac{1}{n} \|\mathbf{K}\| \|\mathbf{K}^{-1}\| \tag{5.6}$$

In the formula: $\|\mathbf{K}\|$ represents the Euclidean norm of the matrix.

In summary, when the stiffness matrix $\mathbf{K}$ is far from the singular boundary, the structure is safer and $\delta_s$ has a larger numerical value. On the other hand, when the stiffness matrix $\mathbf{K}$ is close to the singular boundary, the structure is more dangerous and $\delta_s$ has a smaller numerical value.

### 5.3.2 Physical Meaning of Joint Security Indicators

Suppose that the initial global stiffness matrix $\mathbf{K}$ of the structure is a non-singular $n \times n$ matrix, which can be regarded as a point in the $n \times n$ dimensional space. Taking the two-dimensional space as an example, when the stiffness matrix $\mathbf{K}$ is located in the shaded area in Fig. 5.15, $\mathbf{K}$ is singular, indicating that the structure has failed; when the stiffness matrix $\mathbf{K}$ is located in the blank area in Fig. 5.15, $\mathbf{K}$ is a non-singular matrix, indicating that the structure can continue to bear load; the critical state of structural failure is the boundary between the shaded area and the blank area. Under a given loading pattern, the structure will move along a certain path towards the failure boundary. The distance $d$ from the initial structure to the failure boundary in the given direction can be considered as the safety margin of the structure under the given loading condition. Among them, the shortest distance $d_{min}$ from $\mathbf{K}$ to the failure boundary represents the safety reserve of the structure under the

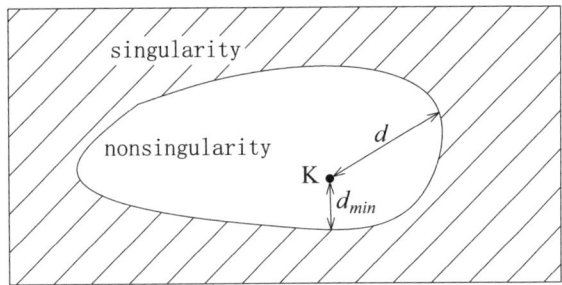

**Fig. 5.15** Distance from initial structure to failure boundary

most unfavorable loading condition, which can represent the overall safety margin of the structure. For example, in the linear eigenvalue buckling analysis process of the shell structure, the uniformly distributed load over the entire span, the half-span uniformly distributed load, etc. represent the loading path from the initial structure to the linear buckling, and the corresponding load amplitude represents the safety reserve of the structure under this loading mode, corresponding to $d$ in Fig. 5.15. Among all the load amplitudes, the smallest corresponds to $d_{min}$ in Fig. 5.15, and the corresponding loading condition is the most unfavorable loading condition. In summary, $d_{min}$ represents the inherent safety performance of the structure, which is mathematically independent of the external load, thus avoiding the shortcoming of the analysis result being load-dependent; at the same time, its structural connotation is the envelope of all loading conditions, which is the safety margin under the most unfavorable loading condition. Mathematically, $d_{min}$ cannot be expressed directly, but it is inversely proportional to the state function $\kappa(\mathbf{K})$ of the matrix, so it can be represented by $\delta_s$.

For the symmetrical plane two-force beam system shown in Fig. 5.16, each beam element's EA/L is a constant, and the range of $\beta$ value is $[0, \pi/2]$. According to the structural concept, when $\beta = 0$, the two-force beam is a geometric instantaneous system without vertical stiffness and cannot bear vertical concentrated force; when $0 < \beta < \pi/4$, the structure begins to have vertical stiffness, and the vertical stiffness gradually increases as $\beta$ increases; when $\beta = \pi/4$, the two beam elements are perpendicular to each other, and the structure has equal stiffness in both perpendicular directions, making it the most stable structure regardless of the direction of the concentrated force; when $\pi/4 < \beta < \pi/2$, the structure's vertical stiffness continues to increase, but the horizontal stiffness gradually decreases; when $\beta = \pi/2$, the two beams overlap, and the horizontal stiffness is zero, and the original structure degrades back into an assembly that cannot bear loads. When $\beta$ varies continuously within $[0, \pi/2]$, the relationship between $\delta_s$ and $\beta$ is calculated, as shown in Fig. 5.17.

From Fig. 5.17, we can conclude that when $\beta = 0$, the bifilar truss is a geometric instantaneous system. Although this geometric instantaneous system can withstand horizontal concentrated forces, the vertical concentrated load is the worst-case scenario, and the system cannot withstand the worst load, so $\delta_s = 0$; when $0 < \beta < \pi/4$, as $\beta$ increases, the structure begins to have vertical stiffness, and its ability to resist vertical loads gradually increases, and $\delta_s$ gradually increases; when

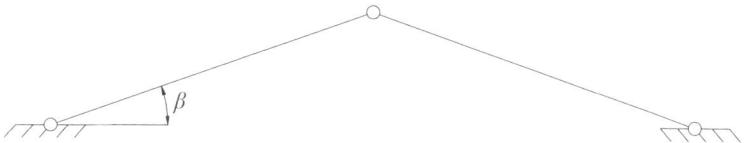

**Fig. 5.16** Plane two-force beam

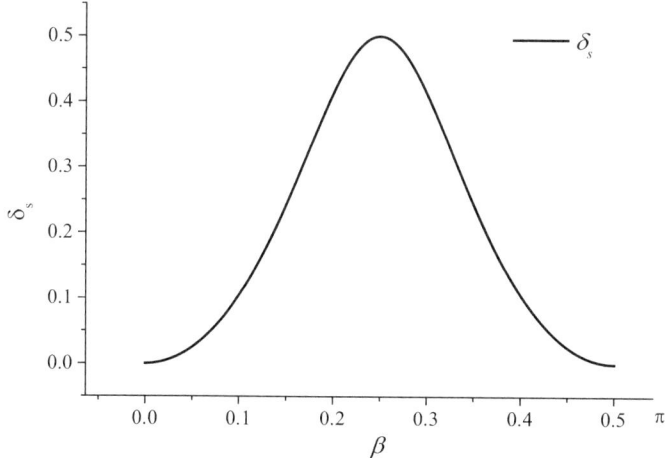

**Fig. 5.17** Relationship between $\delta_s$ and $\beta$

$\beta = \pi/4$, the two members are perpendicular, and the structure has equal stiffness in both perpendicular directions, making it the strongest in resisting loads in all directions, and $\delta_s$ reaches its peak value; when $\pi/4 < \beta < \pi/2$, as $\beta$ increases, the structure's horizontal stiffness gradually decreases, and horizontal loads gradually become the worst-case load scenario, and $\delta_s$ gradually decreases; when $\beta = \pi/2$, the bifilar truss system degrades into an institution, although this institution can withstand vertical concentrated forces, horizontal concentrated loads are the worst-case load scenario, and the system cannot withstand the worst load, so $\delta_s = 0$, and the stiffness matrix is singular, and the calculated $\delta_s$ is also 0. It can be seen that using $\delta_s$ as a structural safety evaluation index not only conforms to the concept of structure, but also automatically identifies the worst load scenario and encompasses various load scenarios, verifying the feasibility of using $\delta_s$ to evaluate the safety of structures. Nafday [30, 31] used this index to evaluate the safety performance of structures and extended its application range to rod-type structures.

### 5.3.3 Topology Optimization Model for Joint Safety Performance

According to Sects. 5.3.1 and 5.3.2, the larger the safety index $\delta_s$ of a joint, the safer the joint is. A secure and reliable joint should rely on the rational distribution of internal materials rather than the massive piling up of materials. Therefore, for a joint with a given upper limit on volume, the topology optimization model for joint safety performance is as follows:

$$\max : \delta_s$$
$$s.t.\ V* - \sum_{i=1}^{N} V_i x_i \geq 0 \tag{5.7}$$
$$x_i = 0\ or\ 1$$

In the formula: $\delta_s$ is the safety index of the joint; $V^*$ is the upper limit of volume, i.e. the target volume of the optimized structure; $V_i$ is the volume of element $i$; $x_i$ is the state function of element $i$, where $x_i = 0$ indicates that the element does not exist, and $x_i = 1$ indicates that the element exists.

### 5.3.4 Sensitivity Coefficient for Units

When a unit is deleted from an existing joint, the change in the entire joint $\delta_s$ is defined as the unit sensitivity coefficient. After deleting the unit from the joint, the safety index of the remaining structure is denoted as $\delta i*\ s$, then the unit's sensitivity $\gamma_{e,i}$ is defined as

$$\gamma_{e,i} = \frac{1}{V_i} \frac{\delta_s}{\delta_s^{i*}} \tag{5.8}$$

In the equation: $V_i$ represents the volume of unit $i$.

If it is a critical unit, deleting the unit will result in a significant decrease in the joint $\delta i*\ s$, therefore the critical unit's sensitivity value is high; if it is an inefficient unit, deleting the unit will result in almost no change in the joint $\delta i*\ s$, therefore the inefficient unit's sensitivity value is low. Among all the unit sensitivities, the maximum unit sensitivity is recorded as $\gamma_{e,max}$:

$$\gamma_{e,max} = \max\{\gamma_{e,1}, \gamma_{e,2}, \ldots, \gamma_{e,i}, \ldots, \gamma_{e,N}\} \tag{5.9}$$

In the formula: N represents the total number of units.

$\gamma_{e,max}$ reflects the distribution of sensitivity of all elements in the structure. If there are individual elements with excessively high sensitivity at a joint, the safety performance of the structure will be significantly reduced once these elements fail, which is not conducive to improving the robustness of the joint. Ideally, the sensitivity of each element in the joint should be distributed evenly.

### 5.3.5 Topology Optimization Algorithm for Joint Safety Performance

The Evolutionary Structural Optimization (ESO) [32] method and its derivative, the Bi-directional Evolutionary Structural Optimization (BESO) [33] method, is a topology optimization algorithm proposed by academician Xie Yiming and others. ESO/BESO obtains an efficient structure by gradually deleting inefficient elements in the structure and retaining efficient elements. Due to its clear structure concept and convenient algorithm implementation, it has been widely used in the engineering [34].

If the unit sensitivity coefficients are used directly for progressive structural optimization, the optimization results will exhibit a checkerboard pattern. The checkerboard pattern is a material distribution pattern with a certain periodicity, thus producing a structure form that is only mathematically feasible but cannot be manufactured in practice. A typical checkerboard pattern is shown in red circles in Fig. 5.18. The reason for the checkerboard pattern is that the unit sensitivity is discontinuous.

To suppress the checkerboard pattern, the filter scheme [33] commonly used in the ESO method is adopted in this paper. First, the element sensitivity calculated by Eq. (5.8) is converted into joint sensitivity according to Eq. (5.10).

$$\gamma_{n,j} = \sum_{i=1}^{M} \omega_i \gamma_{e,i} \quad (5.10)$$

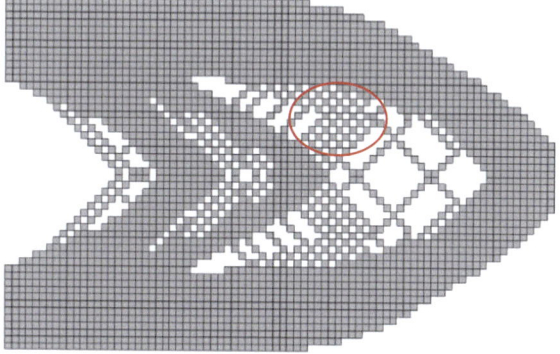

**Fig. 5.18** The checkerboard pattern in the ESO method [33]

In this formula, $\gamma_{n,j}$ represents the sensitivity of joint $j$; $M$ is the number of elements connected to joint $j$; and $\omega_i$ represents the weight of element $i$ connected to joint j, which is calculated according to Eq. (5.11):

$$\omega_i = \frac{1}{M-1}\left(1 - \frac{r_{ij}}{\sum_{i=1}^{M} r_{ij}}\right) \tag{5.11}$$

In the formula: $r_{ij}$ represents the distance from the center of unit i to joint $j$.

After the joint sensitivity is determined by Eqs. (5.10) and (5.11), it can be mapped to elements through a filtering mechanism, generating continuous element sensitivities while preserving the overall distribution pattern. The filtering mechanism should first determine the filtering radius $r_{min}$. Within the region $\Omega_i$ centered at the center of element $i$ and with a radius of $r_{min}$ (as shown in Fig. 5.19), the sensitivity of all joints contributes to the sensitivity of the element, and the closer the joint is to the element, the greater the contribution. The specific calculation of the sensitivity coefficient $\gamma_i$ of element $i$ is shown in Eq. (5.12). At the same time, after the filtering mechanism, the sensitivity coefficients of the real elements adjacent to the unit can also be obtained. After the filtering mechanism, the joint sensitivity is improved in a continuous manner while preserving the distribution pattern, avoiding the checkerboard pattern in subsequent optimization processes.

$$\gamma_i = \frac{\sum_{j=1}^{K} \omega(r_{ij})\gamma_{n,j}}{\sum_{j=1}^{K} \omega(r_{ij})} \tag{5.12}$$

In the formula: K is the total number of joints in $\omega(r_{ij})$ is the weight of joint $j$, as defined in Eq. (5.13):

**Fig. 5.19** Sensitivity coefficients of joint calculation units $i$ within $\Omega_i$

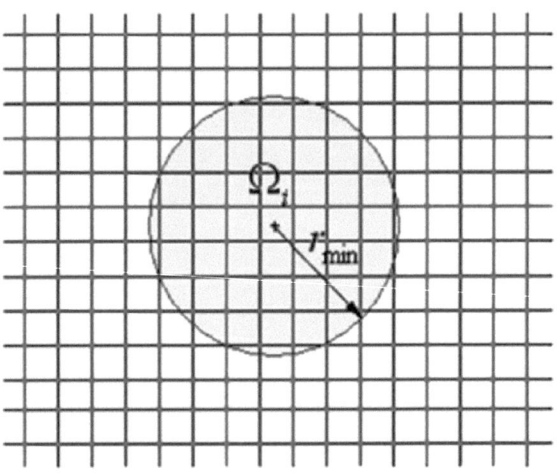

## 5.3 Topology Optimization Based on Safety Performance of Joints

$$\omega(r_{ij}) = r_{\min} - r_{ij} \ (j = 1, 2, \ldots, K-1, K) \tag{5.13}$$

If the current volume of the structure is greater than the target volume, then sort the sensitivities of the filtered elements and remove the *ne_dele* least sensitive solid elements. If after removing inefficient elements, the structure volume is still not less than or equal to the target volume, then halve the optimization step size, i.e. *ne_dele* = *ne_dele*/2. If after removing inefficient elements, calculating $\gamma_{e,\max}$ for joints according to Eq. (5.9) results in $\gamma_{e,\max} > \gamma_{e,threshold}$, threshold, it indicates that there are overly important elements in the current joint; therefore, halve the optimization step size as well, i.e. *ne_dele* = *ne_dele*/2. The first scenario indicates that as optimization progresses, the structure's volume has approached the target volume and thus reducing optimization step size appropriately can yield stable optimization results. The second scenario suggests that in a previous optimization step where an excessively large step size caused significant changes in structural topology; hence reducing this step size is necessary. Additionally, this second scenario effectively sets an upper limit for $\gamma_{e,\max}$ which prevents overly important elements from appearing within the structure and increases its robustness.

If the current structure volume is smaller than the target volume, the sensitivity of the filtered elements is sorted, and *ne_add* entities are added with the highest sensitivity. If adding the high-efficiency element increases the joint volume to be greater than the target volume, the optimization step size is halved, i.e. *ne_add* = *ne_add*/2. As the optimization progresses, the structure will gradually approach the optimal solution. At this time, the optimization step size should be reduced to make the optimization results converge faster to the optimal solution and obtain stable optimization results. The optimization example shows that the adaptive step size strategy can improve the convergence rate of the optimization and obtain stable and reliable optimization results. When the progressive structure optimization method evolves to a sufficient number of generations, the optimization stops. The joint topology optimization algorithm based on safety performance is shown in Fig. 5.20.

### 5.3.6 Optimized Joints Based on Security Performance

For example, as shown in Fig. 5.21, the shaded diagonal area represents the beam elements, and there are three beam elements connected to the two-dimensional joint. The mesh area is the joint design domain, with a size of 24 × 32. The initial joint occupies all the mesh cells, and the initial volume $V_0 = 768$.

With the optimization model given by Eq. (5.7), the target volume $V^* = 274$. In the optimization algorithm, the initial step sizes *ne_dele* = 8, *ne_add* = 8, $\gamma_{e,threshold}$ = 1.2. After 180 optimization steps, the joints evolve to the optimized solution. The evolutionary process of the joints is shown in Fig. 5.22, and the final optimized joint form is shown in Fig. 5.22d.

To reflect the contribution of unit volume to joint safety performance, the unit volume safety factor $\delta vs$ is defined, as shown in Eq. (5.14). The joint safety factor $\delta_s$

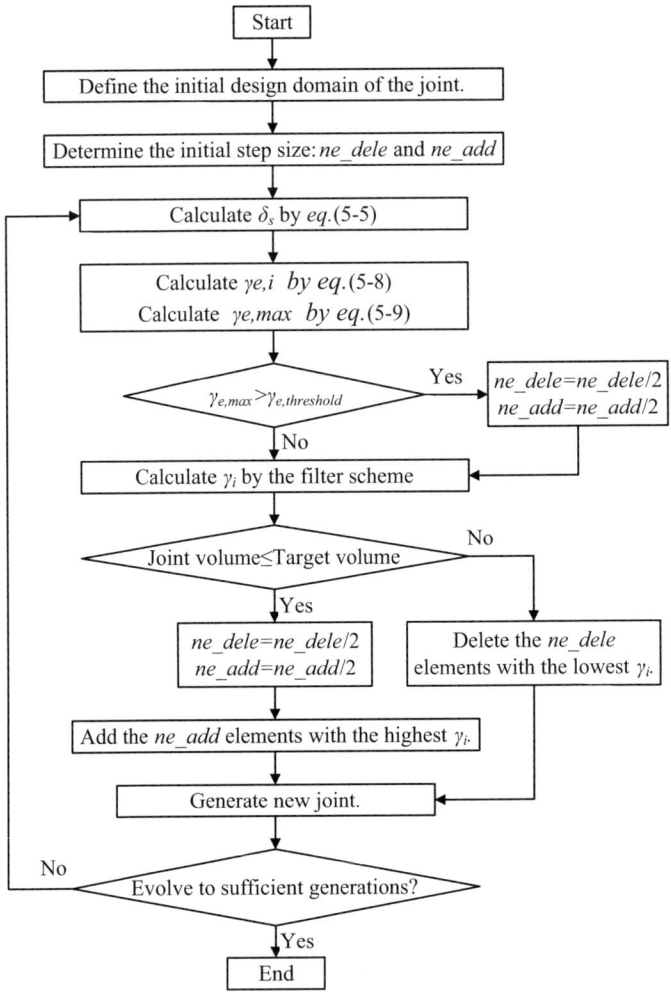

**Fig. 5.20** Optimization algorithm for joint topology based on security performance

can reflect the overall safety performance of the joint, but cannot distinguish whether its safety performance is due to the massive piling of materials or due to the rational distribution of materials. The unit volume safety factor $\delta v\ s$, however, can distinguish this. The higher its value, the more reasonable the material distribution is.

$$\delta_s^v = \frac{1}{V}\delta_s \tag{5.14}$$

When $V^* = 274$, the optimization process of $\delta_s$ and $\delta v\ s$ is shown in Fig. 5.23. From Fig. 5.23, we can see that: at the beginning of the optimization, as inefficient units are deleted and material distribution becomes more reasonable, $\delta_s$ gradually

5.3 Topology Optimization Based on Safety Performance of Joints

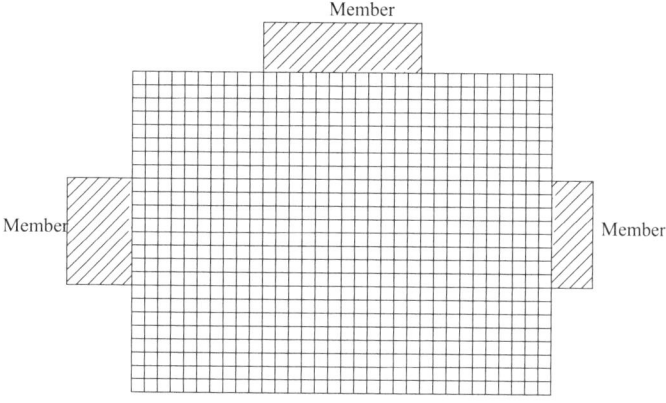

**Fig. 5.21** 2D joint initial design domain

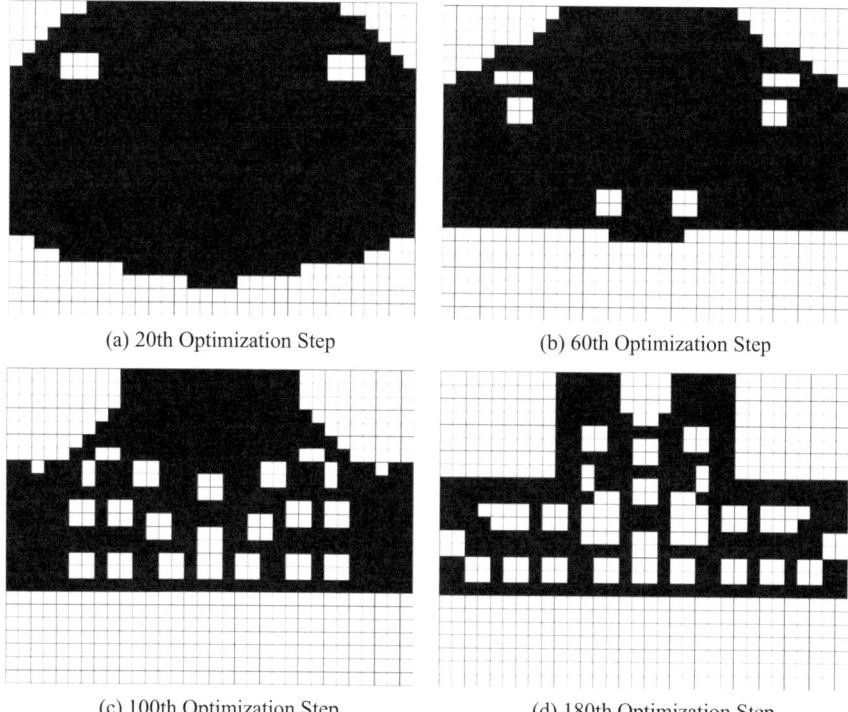

(a) 20th Optimization Step

(b) 60th Optimization Step

(c) 100th Optimization Step

(d) 180th Optimization Step

**Fig. 5.22** Evolution of joints when $V^* = 274$

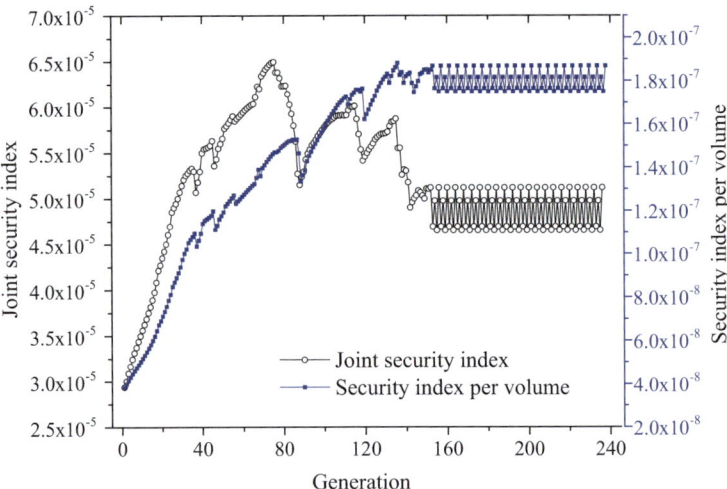

**Fig. 5.23** The optimization process of $\delta_s$ and $\delta v\ s$ at $V^* = 274$

increases; later, due to the volume limitation, $\delta_s$ decreases slightly, but the safety factor per unit volume, $\delta v\ s$, continues to increase; after 160 optimization steps, $\delta_s$ and $\delta v\ s$ both fluctuate within a limited range periodically, indicating that the structure has evolved to an optimized solution.

The sensitivity distribution of the initial joint is shown in Fig. 5.24, and the unit sensitivity distribution of the optimized joint is shown in Fig. 5.25. The initial joint occupies the entire design domain, and joint failure generally does not occur within the joint. The failure location is generally located at the connection between the joint and the member. In Fig. 5.24, individual highly sensitive units are distributed at the connection between members, and are clearly higher than the internal units. This is consistent with the structural concept, further verifying the safety assessment method proposed in this chapter. Comparing Figs. 5.24 and 5.25 can be found that the optimized joint not only has a more streamlined structural form, but also some important units are distributed in the middle and lower part of the joint. The unit sensitivity distribution of the optimized joint is more uniform than that of the initial joint. However, the unit sensitivity is low in the upper part of the optimized joint, indicating that there is still room for further optimization of the joint.

In the second joint safety performance optimization example, the target volume $V^* = 254$ is slightly smaller than the first example in order to expect a more reasonable joint form and element sensitivity distribution. At the same time, in order to test the effectiveness of the adaptive optimization step size strategy, the initial optimization step size $ne\_dele = 4$, $ne\_add = 4$ is used in this example, and other optimization parameters are the same as the first example. The evolution of joint in the optimization process is shown in Fig. 5.26, and the final optimized joint is shown in Fig. 5.26d.

When $V^* = 254$, the optimization process of $\delta_s$ and $\delta v\ s$ is shown in Fig. 5.27. From Fig. 5.27, we can see that: in the initial stage of optimization, as inefficient

## 5.3 Topology Optimization Based on Safety Performance of Joints

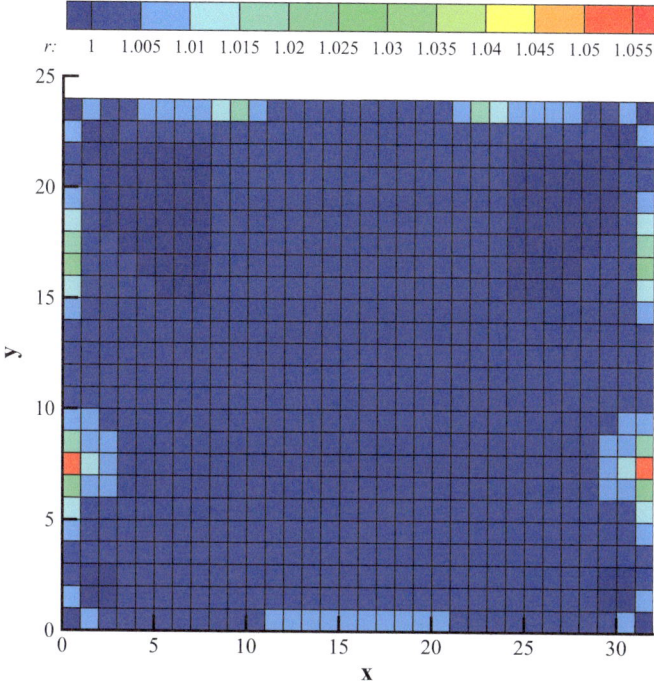

**Fig. 5.24** Sensitivity distribution of elements of the initial joint

units are deleted and material distribution becomes more reasonable, $\delta_s$ gradually increases; later, due to the volume limitation, $\delta_s$ decreases slightly, but the safety factor per unit volume $\delta v\ s$ continues to increase; after 240 optimization steps, $\delta_s$ and $\delta v\ s$ both fluctuate within a limited range periodically, indicating that the structure has evolved to an optimized solution.

The distribution of element sensitivity at the optimized joint is shown in Fig. 5.28. Compared with the sensitivity distribution of the corresponding joint at $V^* = 274$ (shown in Fig. 5.25), the sensitivity distribution at the optimized joint is more evenly distributed. The sensitive elements are distributed in the upper, middle, and lower parts of the joint. At the same time, the element sensitivity distribution in the entire joint is more evenly distributed, indicating that the element distribution is in a reasonable position and can effectively improve the safety performance of the joint.

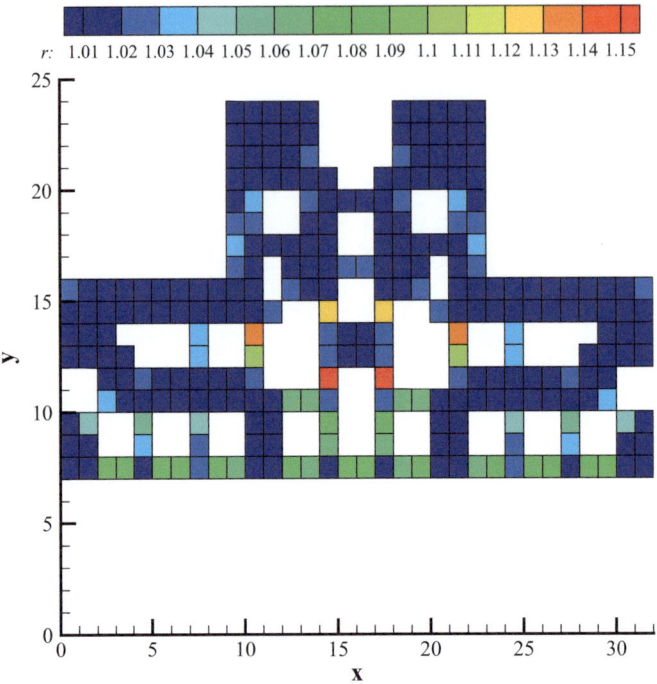

**Fig. 5.25** Sensitivity distribution of elements at the optimized joint with V* = 274

## 5.4 Chapter Summary

(1) To improve the rotational stiffness of the joints, separate topological optimization models were established for maximum rotational stiffness joints with a fixed mass and minimum mass with a fixed stiffness. In the topological optimization model for maximum rotational stiffness with a fixed mass, the target is to maximize the sum of the joint rotational stiffnesses under the premise of giving a upper limit of joint mass and stress. In the topological optimization model for minimum mass with a fixed stiffness, the target is to minimize the joint mass under the premise of giving a lower limit of joint stiffness and stress. In the two topological optimization models, the joint stiffness is represented by the self-balancing force system to avoid the dependency of the optimization result on the loading conditions and structure form, making the optimized joints universally applicable. In the topological optimization process, the equivalent concentrated load due to inertia force is applied to the joints to avoid the stress singularity problem caused by the direct application of concentrated force on the design domain and the disadvantage of joint form limitation. It also avoids the provision of balancing force by the support, making the boundary condition only limit the translational movement of the rigid body, thus obtaining independent

## 5.4 Chapter Summary

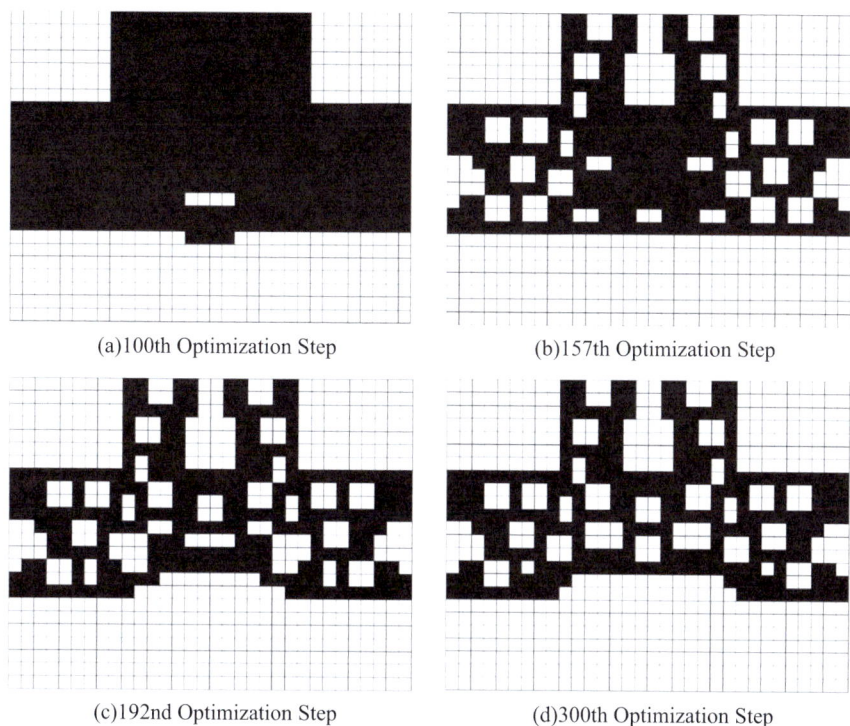

(a) 100th Optimization Step  (b) 157th Optimization Step

(c) 192nd Optimization Step  (d) 300th Optimization Step

**Fig. 5.26** Evolution of joints when $V^* = 254$

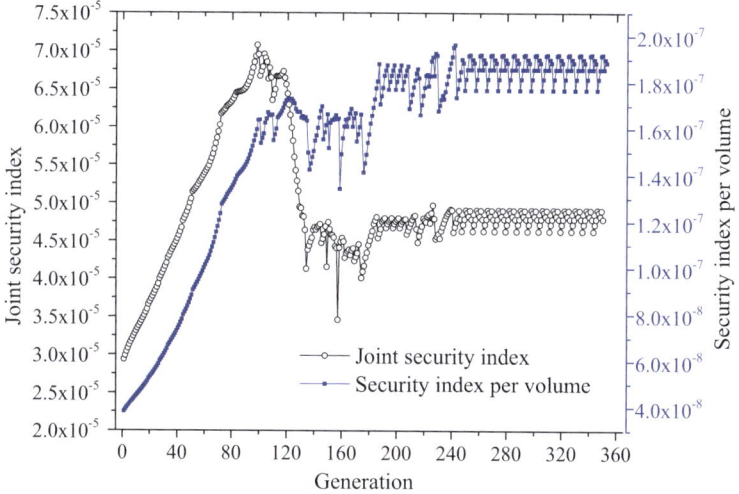

**Fig. 5.27** The optimization process of $\delta_s$ and $\delta v\ s$ at $V^* = 254$

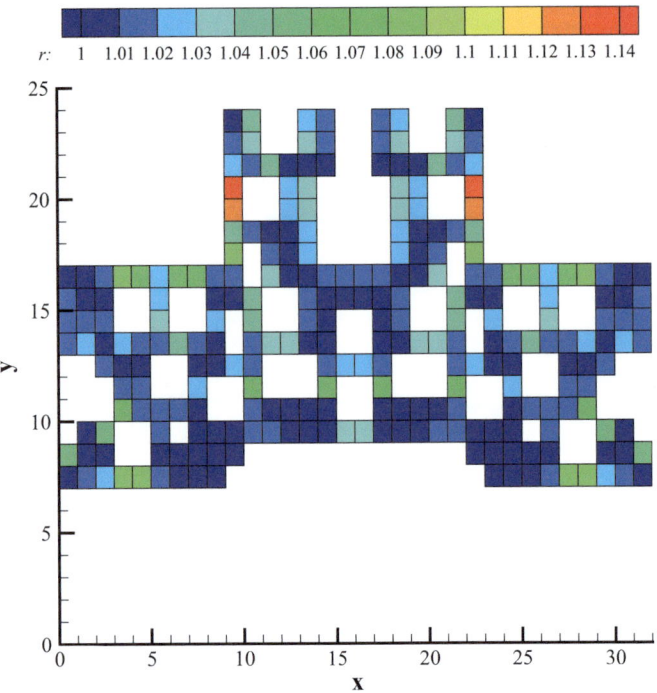

**Fig. 5.28** Sensitivity distribution of elements at the optimized joint with V* = 254

of the boundary condition optimization result. Through the above methods, the key technical difficulties in topological optimization were solved. For the optimized joints, the spherical joint ends were designed by combining the structure. The spherical joint end allows the joint to have directionless, good geometric adaptability, good member adaptability, and auxiliary alignment function. The joint designed by combining topology optimization with construction method has better mechanical performance with less weight than the traditional welded hollow ball joint.

(2) To improve the safety performance of joints, an independent structural safety evaluation index was first proposed, which is not dependent on the load. This index focuses on the inherent properties of the structure and measures the safety margin of the structure from the initial stiffness matrix to the singular edge (i.e., the failure edge) to evaluate the structural safety. The index is mathematically independent of the load and is expressed in a concise manner, with its engineering significance encompassing various loading conditions and the safety margin of the joint under the most unfavorable loading mode. The rationality of the index was verified by a simple two-force beam structure. Then, the safety performance of joints was quantitatively expressed by the index, and the optimization goal was to improve the safety performance of joints,

with the constraint of the joint volume. A topology optimization model for joint safety performance was established, and an optimization algorithm was formulated accordingly. The corresponding optimization model was solved. Finally, topology optimization was performed on two two-dimensional joints, resulting in a simplified and compact structure with high safety index, verifying the feasibility of the optimization algorithm.

# References

1. Liu X (1984) The application and development of West German lattice structure and its joints. In: Proceedings of the second space structure academic exchange conference (vol II)
2. Liu X (2000) Review of joints of space structures at home and abroad. In: Proceedings of the ninth space structure academic conference, Xiaoshan
3. Yin D, Liu S (1994) Review of joint types of steel lattice shell structure. Steel Struct 2:115–127
4. Hanaor A (1995) Characteristics of prefabricated spatial frame systems. Int J Space Struct 10(3):151–173
5. Weng Z, Zhao Y, Jin Y et al (2018) Classification and development demand of prefabricated joints for spatial grid structures. J Build Struct 39(3):32–38
6. Arciszewski T, Uduma K (1988) Shaping of spherical joints in space structures. Int J Space Struct 3(3):171–183
7. Bangash MYH, Bangash T (2003) Elements of spatial structures: analysis and design. Thomas Telford, pp 463–437
8. Makowski ZS (2002) Development of jointing systems for modular prefabricated steel space structures. In: Proceedings of the international symposium. IASS, Warsaw, pp 17–41
9. Liu XL (2000) Review of spatial structure joints at home and abroad. In: Proceedings of the 9th academic conference on spatial structures. Bridge and Structural Engineering Branch of Chinese Society of Civil Engineering, Xiaoshan, pp 10–18
10. Luo Y, Shen Z (1995) Influence of joints on load-bearing performance of lattice shells. J Tongji Univ (Nat Sci) 23(1):21–25
11. Liu H, Luo Y, Xu X (2013) Influence of welded ball joint stiffness on finite element analysis accuracy of lattice shells. Eng Mechan 30(01):350–358+364
12. Cao Z, Fan F, Ma M, Wang W (2010) Test of bolted ball joints and its application in single-layer lattice shells. J Harb Inst Technol 42(04):525–530
13. Fan F, Ma H, Cao Z et al (2011) A new classification system for the joints used in lattice shells. Thin-Walled Struct 49(12):1544–1553
14. Nguyen PC, Kim SE (2014) Nonlinear inelastic time-history analysis of three-dimensional semi-rigid steel frames. J Constr Steel Res 101:192–206
15. Han Q, Liu Y, Zhang J et al (2017) Mechanical behaviors of the assembled Hub (AH) joints subjected to bending moment. J Constr Steel Res 138:806–822
16. Feng R, Liu F, Yan G et al (2017) Mechanical behavior of ring-sleeve joints of single-layer reticulated shells. J Constr Steel Res 128:601–610
17. Feng R, Wang X, Chen Y et al (2018) Static performance of double-ring joints for freeform single-layer grid shells subjected to a bending moment and shear force. Thin-Walled Struct 131:135–150
18. Fan F, Ma M, Ma Y (2019) Research progress and key issues of semi-rigid joint lattice shells. Eng Mechan 36(7):1–7, 29
19. Han Q, Wang C, Xu Y et al (2020) Mechanical performance of AH joints and influence on the stability behaviour of single-layer cylindrical shells. Thin-Walled Struct 138:106459

20. Bendsøe MP (1989) Optimal shape design as a material distribution problem. Struct Optim 1(4):193–202
21. Duysinx P, Bendsøe MP (1998) Topology optimization of continuum structures with local stress constraints. Int J Numer Meth Eng 43(8):1453–1478
22. Sigmund O, Maute K (2013) Topology optimization approaches. Struct Multidiscip Optim 48(6):1031–1055
23. Industrial standard of the People's Republic of China (2010) JGJ7-2010 technical specification for space grid structure. China Architecture and Building Press, Beijing
24. Frangopol DM, Iizuka M, Yoshida K (1992) Redundancy measures for design and evaluation of structural systems. J Offshore Mech Arct Eng 114(4):285–290
25. Brett C, Lu Y (2013) Assessment of robustness of structures: current state of research. Front Struct Civ Eng 7(4):356–368
26. Starossek U, Haberland M (2011) Approaches to measures of structural robustness. Struct Infrastruct Eng 7(7–8):625–631
27. Wu X, Blockley DI, Woodman NJ (1993) Vulnerability of structural systems Part 1: rings and clusters. Civ Eng Syst 10(4):301–317
28. Wu X, Blockley DI, Woodman NJ (1993) Vulnerability of structural systems Part 2: failure scenarios. Civ Eng Syst 10(4):319–333
29. Demmel JW (1987) On condition numbers and the distance to the nearest ill-posed problem. Numer Math 51(3):251–289
30. Nafday AM (2008) System safety performance metrics for skeletal structures. J Struct Eng 134(3):499–504
31. Nafday AM (2011) Consequence-based structural design approach for black swan events. Struct Saf 33(1):108–114
32. Xie YM, Steven GP (1993) A simple evolutionary procedure for structural optimization. Comput Struct 49(5):885–896
33. Huang X, Xie M (2010) Evolutionary topology optimization of continuum structures: methods and applications. Wiley
34. Xia L, Xia Q, Huang X et al (2018) Bi-directional evolutionary structural optimization on advanced structures and materials: a comprehensive review. Archiv Comput Methods Eng 25(2):437–478

**Open Access** This chapter is licensed under the terms of the Creative Commons Attribution-NonCommercial-NoDerivatives 4.0 International License (http://creativecommons.org/licenses/by-nc-nd/4.0/), which permits any noncommercial use, sharing, distribution and reproduction in any medium or format, as long as you give appropriate credit to the original author(s) and the source, provide a link to the Creative Commons license and indicate if you modified the licensed material. You do not have permission under this license to share adapted material derived from this chapter or parts of it.

The images or other third party material in this chapter are included in the chapter's Creative Commons license, unless indicated otherwise in a credit line to the material. If material is not included in the chapter's Creative Commons license and your intended use is not permitted by statutory regulation or exceeds the permitted use, you will need to obtain permission directly from the copyright holder.

# Chapter 6
# Optimization of Stability of Single-Layer Gridshells Considering Joint Stiffness

**Abstract** The members and joints are the two major components of a single-layer gridshell, and are also important factors that significantly affect the stability performance of the structure. With the development of the construction industrialization in China, scholars have developed many semi-rigid new types of joints. Compared with traditional joints, these new joints are neither ideal rigidly connected nor ideal hinged. However, in the current design method of single-layer gridshells, it is still assumed that the joints are ideal rigidly connected first, and the member design is carried out. Then, the stability calculation is based on the assumption of rigid joints. This two-stage design method not only disconnects the mutual influence between members and joints, but also restricts the use and promotion of new types of joints. Related research results have shown that the stiffness of the joints has a significant impact on the stability performance of the gridshell. Therefore, there is still room for improvement in the current design method. This chapter further expands the theory of shape fragility, introduces joint stiffness into the theory of shape fragility, defines the relative change gradient of shape degree of freedom $gra\_r$ for semi-rigid joints, and reveals the instability mechanism of single-layer shell structures from both the beam and joint levels. Then, taking a single-layer shell structure as an example, the chapter quantitatively examines the influence of joint stiffness on the stability performance of the shell structure from the perspectives of structural stable bearing capacity and joint shape degree of freedom. It determines the reasonable range of joint stiffness suitable for design. To consider the influence of both the joint and the beam on the stability of the shell structure at the same time, this chapter proposes a stable optimization design method for single-layer shell structures considering joint stiffness. The optimization model takes both the beam and the joint as optimization variables, maximizes the minimum value of the relative change gradient of joint shape degree of freedom $gra\_r_{min}$ as the optimization goal, considers the constraints of steel usage for beams, joints, beam design limits, etc., and develops corresponding optimization algorithms. Three single-layer shell structures with different spans are used as examples to verify the stable optimization design method considering joint stiffness. The stable optimization results of three examples show that the stable optimization design method considering joint stiffness can optimize the distribution of beam section and joint stiffness under the premise that the steel used at joints is significantly lower than that of the traditional design method, thereby improving the stability bearing capacity of single-layer shell structure. Furthermore, the method verifies the anti-collapse performance of semi-rigid joint optimized shell structure, and the calculation results show that the maximum displacement response of semi-rigid joint optimized shell structure in a 9° rare earthquake is basically the same as that of the rigid joint optimized shell structure, and its collapse critical seismic acceleration peak value is significantly higher than the peak value of the 9° rare earthquake seismic acceleration specified in the seismic code.

## 6.1 Damage Theory Considering Joint Stiffness

Chapter 2 builds on the classical theory of geometric fragility, introducing the geometric stiffness matrix to consider external factors such as loads, supports, and constraints, thereby expanding the scope of the theory. Based on the assumption of rigid joints, it reveals the instability mechanism of gridshells from the perspective of degradation of joint geometry. However, many new joints are not ideal rigidly connected joints, but semi-rigid joints between rigidly connected and hinged joints. This section first reviews the stiffness matrix of beam elements considering joint stiffness, then introduces joint stiffness into the theory of geometric fragility, thereby revealing the instability mechanism of single-layer gridshells with semi-rigid joints.

### 6.1.1 Modifying the Stiffness Matrix of a Rectangular Element

Monforton [1] derived the stiffness matrix for beam elements with elastic joints. As shown in Fig. 6.1, the beam element with semi-rigid joints has a length of L, a cross-sectional area of A, and moments of inertia about the $x$, $y$, and $z$ axes of $J$, $I_Y$, and $I_Z$, respectively. At joint i, the elastic joint has rotational stiffness about the $y$ and $z$ axes of kyi and kzi, respectively. At joint $j$, the elastic joint has rotational stiffness about the $y$ and $z$ axes of $k_{yj}$ and $k_{zj}$, respectively. By modifying the classical Euler beam element stiffness matrix to account for the influence of joint stiffness, the modified element stiffness matrix $\mathbf{K_e}$ is expressed as Equation.

$$\mathbf{K_e} = \begin{bmatrix} \mathbf{k_{ii}} & \mathbf{k_{ij}} \\ \mathbf{k_{ji}} & \mathbf{k_{jj}} \end{bmatrix} \tag{6.1}$$

In the equation: $\mathbf{k_{ii}}$, $\mathbf{k_{ij}}$, $\mathbf{k_j}$, and $\mathbf{k_{jj}}$ are respectively given by Equations to

## 6.1 Damage Theory Considering Joint Stiffness

**Fig. 6.1** Beam elements with semi-rigid joints

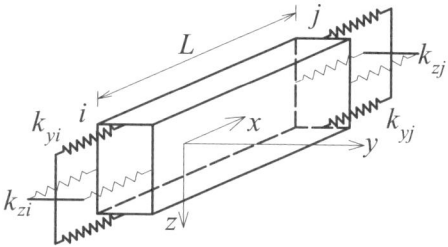

$$\mathbf{k_{ii}} = E \begin{bmatrix} \frac{A}{L} & 0 & 0 & 0 & 0 & 0 \\ 0 & \frac{\gamma_{zi}+\gamma_{zj}+\gamma_{zi}\gamma_{zj}}{4-\gamma_{zi}\gamma_{zj}}\frac{12I_Z}{L^3} & 0 & 0 & 0 & \frac{\gamma_{zi}(2+\gamma_{zj})}{4-\gamma_{zi}\gamma_{zj}}\frac{6I_Z}{L^2} \\ 0 & 0 & \frac{\gamma_{yi}+\gamma_{yj}+\gamma_{yi}\gamma_{yj}}{4-\gamma_{yi}\gamma_{yj}}\frac{12I_Y}{L^3} & 0 & -\frac{\gamma_{yi}(2+\gamma_{yj})}{4-\gamma_{yi}\gamma_{yj}}\frac{6I_Y}{L^2} & 0 \\ 0 & 0 & 0 & \frac{GJ}{EL} & 0 & 0 \\ 0 & 0 & -\frac{\gamma_{yi}(2+\gamma_{yj})}{4-\gamma_{yi}\gamma_{yj}}\frac{6I_Y}{L^2} & 0 & \frac{3\gamma_{yi}}{4-\gamma_{yi}\gamma_{yj}}\frac{4I_Y}{L} & 0 \\ 0 & \frac{\gamma_{zi}(2+\gamma_{zj})}{4-\gamma_{zi}\gamma_{zj}}\frac{6I_Z}{L^2} & 0 & 0 & 0 & \frac{3\gamma_{zi}}{4-\gamma_{zi}\gamma_{zj}}\frac{4I_Z}{L} \end{bmatrix} \quad (6.2)$$

$$\mathbf{k_{ij}} = E \begin{bmatrix} -\frac{A}{L} & 0 & 0 & 0 & 0 & 0 \\ 0 & \frac{\gamma_{zi}+\gamma_{zj}+\gamma_{zi}\gamma_{zj}}{\gamma_{zi}\gamma_{zj}-4}\frac{12I_Z}{L^3} & 0 & 0 & 0 & \frac{\gamma_{zj}(2+\gamma_{zi})}{4-\gamma_{zi}\gamma_{zj}}\frac{6I_Z}{L^2} \\ 0 & 0 & \frac{\gamma_{yi}+\gamma_{yj}+\gamma_{yi}\gamma_{yj}}{\gamma_{yi}\gamma_{yj}-4}\frac{12I_Y}{L^3} & 0 & \frac{\gamma_{yj}(2+\gamma_{yi})}{\gamma_{yi}\gamma_{yj}-4}\frac{6I_Y}{L^2} & 0 \\ 0 & 0 & 0 & -\frac{GJ}{EL} & 0 & 0 \\ 0 & 0 & \frac{\gamma_{yj}(2+\gamma_{yi})}{4-\gamma_{yi}\gamma_{yj}}\frac{6I_Y}{L^2} & 0 & \frac{3\gamma_{yi}\gamma_{yj}}{4-\gamma_{yi}\gamma_{yj}}\frac{2I_Y}{L} & 0 \\ 0 & \frac{\gamma_{zi}(2+\gamma_{zj})}{\gamma_{zi}\gamma_{zj}-4}\frac{6I_Z}{L^2} & 0 & 0 & 0 & \frac{3\gamma_{zi}\gamma_{zj}}{4-\gamma_{zi}\gamma_{zj}}\frac{2I_Z}{L^2} \end{bmatrix} \quad (6.3)$$

$$\mathbf{k_{ji}} = E \begin{bmatrix} -\frac{A}{L} & 0 & 0 & 0 & 0 & 0 \\ 0 & \frac{\gamma_{zi}+\gamma_{zj}+\gamma_{zi}\gamma_{zj}}{\gamma_{zi}\gamma_{zj}-4}\frac{12I_Z}{L^3} & 0 & 0 & 0 & \frac{\gamma_{zi}(2+\gamma_{zj})}{\gamma_{zi}\gamma_{zj}-4}\frac{6I_Z}{L^2} \\ 0 & 0 & \frac{\gamma_{yi}+\gamma_{yj}+\gamma_{yi}\gamma_{yj}}{\gamma_{yi}\gamma_{yj}-4}\frac{12I_Y}{L^3} & 0 & \frac{\gamma_{yi}(2+\gamma_{yj})}{4-\gamma_{yi}\gamma_{yj}}\frac{6I_Y}{L^2} & 0 \\ 0 & 0 & 0 & -\frac{GJ}{EL} & 0 & 0 \\ 0 & 0 & \frac{\gamma_{yj}(2+\gamma_{yi})}{\gamma_{yi}\gamma_{yj}-4}\frac{6I_Y}{L^2} & 0 & \frac{3\gamma_{yi}\gamma_{yj}}{4-\gamma_{yi}\gamma_{yj}}\frac{2I_Y}{L} & 0 \\ 0 & \frac{\gamma_{zj}(2+\gamma_{zi})}{4-\gamma_{zi}\gamma_{zj}}\frac{6I_Z}{L^2} & 0 & 0 & 0 & \frac{3\gamma_{zi}\gamma_{zj}}{4-\gamma_{zi}\gamma_{zj}}\frac{2I_Z}{L^2} \end{bmatrix} \quad (6.4)$$

$$\mathbf{k_{jj}} = E \begin{bmatrix} \frac{A}{L} & 0 & 0 & 0 & 0 & 0 \\ 0 & \frac{\gamma_{zi}+\gamma_{zj}+\gamma_{zi}\gamma_{zj}}{4-\gamma_{zi}\gamma_{zj}}\frac{12I_Z}{L^3} & 0 & 0 & 0 & \frac{\gamma_{zj}(2+\gamma_{zi})}{4-\gamma_{zi}\gamma_{zj}}\frac{-6I_Z}{L^2} \\ 0 & 0 & \frac{\gamma_{yi}+\gamma_{yj}+\gamma_{yi}\gamma_{yj}}{4-\gamma_{yi}\gamma_{yj}}\frac{12I_Y}{L^3} & 0 & \frac{\gamma_{yj}(2+\gamma_{yi})}{4-\gamma_{yi}\gamma_{yj}}\frac{6I_Y}{L^2} & 0 \\ 0 & 0 & 0 & \frac{GJ}{EL} & 0 & 0 \\ 0 & 0 & \frac{\gamma_{yj}(2+\gamma_{yi})}{4-\gamma_{yi}\gamma_{yj}}\frac{6I_Y}{L^2} & 0 & \frac{3\gamma_{yj}}{4-\gamma_{yi}\gamma_{yj}}\frac{4I_Y}{L} & 0 \\ 0 & \frac{\gamma_{zj}(2+\gamma_{zi})}{4-\gamma_{zi}\gamma_{zj}}\frac{-6I_Z}{L^2} & 0 & 0 & 0 & \frac{3\gamma_{zj}}{4-\gamma_{zi}\gamma_{zj}}\frac{4I_Z}{L} \end{bmatrix} \quad (6.5)$$

In the equation: $\gamma_{yi} = \dfrac{1}{1+3EI_Y/Lk_{yi}}$; $\gamma_{yj} = \dfrac{1}{1+3EI_Y/Lk_{yj}}$; $\gamma_{zi} = \dfrac{1}{1+3EI_Z/Lk_{zi}}$; $\gamma_{zj} = \dfrac{1}{1+3EI_Z/Lk_{zj}}$.

### 6.1.2 The Degree of Freedom in the Joint Configuration Considering the Joint Stiffness

For a network shell with $n$ non-restrained joints, the assumption that the joints are ideal rigidly connected joints is no longer made, but rather elastic joints. The overall stiffness matrix $\mathbf{K}$ is shown in Equation.

$$\mathbf{K} = \sum \mathbf{K_e} = \begin{bmatrix} \mathbf{K_{11}} & \cdots & & & \mathbf{K_{1n}} \\ & \ddots & & & \\ \vdots & & \mathbf{K_{kk}} & & \vdots \\ & & & \ddots & \\ \mathbf{K_{n1}} & & & \cdots & \mathbf{K_{nn}} \end{bmatrix} \quad (6.6)$$

In this equation, $\mathbf{K_e}$ represents the element stiffness matrix that takes into account the joint stiffness, as shown in Equation; $\mathbf{K_{kk}}$ is the sub-stiffness matrix of the rigid matrix $\mathbf{K}$ that is related to joint $k$, with the dimension C equal to the number of degrees of freedom of the joint.

Because the joint stiffness is taken into account in the element stiffness matrix $\mathbf{K_e}$, the overall stiffness matrix $\mathbf{K}$ also reflects the contribution of both the member stiffness and the joint stiffness to the overall stiffness of the structure. Similar to the initial configuration stiffness of rigid joints discussed in Chap. 2, the initial joint configuration stiffness considering joint stiffness can be calculated according to Equation.

$$q_{k,0} = \det(\mathbf{K}_{kk}) \quad (6.7)$$

In the equation: $\mathbf{K_{kk}}$ as shown in Equation.

Under the action of loads, the geometric stiffness matrix $\mathbf{K_{GC}}$ integrated with the compression member is introduced into the nodal well-formedness, and the elastic nodal well-formedness $q_{k,1}$ of joint $k$ under load is defined as shown in Equation.

$$q_{k,1} = det(\mathbf{K}_{kk} + \mathbf{K}_{GCkk}) \quad (6.8)$$

In the equation: $\mathbf{K_{kk}}$ is defined in Equation; $\mathbf{K_{GCkk}}$ is the sub-stiffness matrix related to joint $k$ in the $\mathbf{K_{GC}}$; $\mathbf{K_{GC}}$ is defined in Equation.

$q_{k,0}$ is a comprehensive measure of the initial stiffness of semi-rigid joint $k$; $q_{k,1}$ is a measure of the stiffness of semi-rigid joint $k$ under load and other external factors. The relative change in shape degree of semi-rigid joint $k$ under load is defined as $gra\_r_k$, which is used to measure the degree of stiffness degradation of semi-rigid joints, as shown in Equation.

$$gra\_r_k = \frac{q_{k,1} - q_{k,0}}{q_{k,0}} \quad (6.9)$$

In the equation: $q_{k,0}$ and $q_{k,1}$ are respectively given in equations and.

### 6.1.3 The Mechanism of Instability in Mesh Structures with Consideration of Joint Stiffness

From equations to, it can be concluded that in the overall structure, the stiffness of members (EA, EI, etc.) and the rotational stiffness of joints ($k_y$, $k_z$, etc.) jointly affect the deformation degree of joints. The classical theory of geometric fragility assumes a rigid joint and only considers the contribution of member stiffness to the deformation degree of joints. The geometric fragility theory expanded in this chapter can reveal the instability mechanism of the network shell structure from both the member and joint perspectives, from the perspective of joint stiffness degradation: (1) Due to the compression of members, the related joint stiffness degrades significantly, with $gra\_r$ far lower than other joints. When the load is increased until even more joints lose their ability to resist the load, the structure loses stability. Among the joints with the smallest relative change in deformation gradient ($gra\_r_{min}$), the degradation of stiffness is most significant, and they are prone to instability during the loading process, being the key joints determining the stability of the network shell; (2) The lower the $gra\_r_{min}$ of the network shell, the more significant the stiffness degradation, and the more obvious the trend of losing stability. The lower the stable bearing capacity, the lower the stability, and vice versa.

## 6.2 The Influence of Joint Rotational Stiffness on the Stability Performance of Single-Layer Gridshells

### 6.2.1 Dimensionless Torsional Stiffness and Stable Bearing Capacity

Han et al. [2] established a three-dimensional mechanical model of spatial structure joints, as shown in Fig. 6.2. Under the force system My (show in Fig. 5.5), the out-of-plane strain energy $C_{My}$ of the joint can be calculated according to Equation:

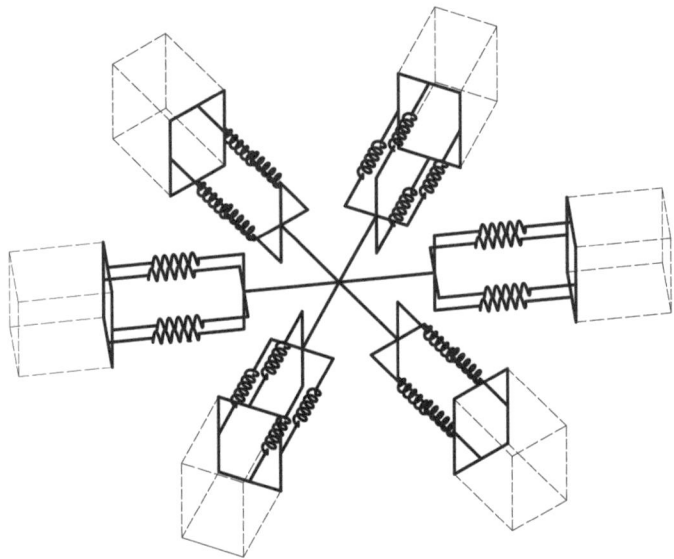

**Fig. 6.2** 3D mechanical model of joint structures

$$6 \times \frac{1}{2} k_y \left(\frac{M}{k_y}\right)^2 = C_{My} \tag{6.10}$$

In the equation, $M$ represents the unit out-of-plane moment applied at the joint end, as shown in Fig. 5.6.

From Equations, the rotational stiffness $k_y$ of the joint in the plane perpendicular to the shell can be calculated. Similarly, the rotational stiffness $k_z$ of the joint in the plane of the shell can be calculated from the joint's mechanical model and the strain energy $C_{Mz}$ in the plane of the shell according to Equation.

$$6 \times \frac{1}{2} k_z \left(\frac{M}{k_z}\right)^2 = C_{Mz} \tag{6.11}$$

In the equation, $M$ represents the unit bending moment applied to the joint end in the plane, as shown in Fig. 5.6.

Unlike joints in frame structures, the bending stiffness in spatial structures in both directions is not significantly different. Assuming that the bending stiffness in both directions is equal, i.e. $k_y = k_z = k_m$. The ability of the joint to constrain the angle of the member is not only related to the joint's own stiffness ($k_y$, $k_z$) but also related to the linear bending stiffness of the connected member. At the same time, in order to generalize the specific conclusion obtained from a single case to a general conclusion, the dimensionless treatment of the joint stiffness is carried out, and the dimensionless quantity $\alpha$ represents the joint stiffness.

## 6.2 The Influence of Joint Rotational Stiffness on the Stability Performance …

$$\alpha = \frac{k_m}{EI/L} \qquad (6.12)$$

In the equation: $k_m$ is the rotational stiffness of the joint; $E$ is the modulus of elasticity; $I$ is the moment of inertia of the main steel pipe; $L$ is the geometric length of the main steel pipe. $EI/L$ is the line bending stiffness of the main steel pipe.

When the joint is an ideal rigid joint, the maximum elastic stable bearing capacity of the structure is $P_{rigid\ cr}$. When the joint stiffness is $\alpha$, the corresponding structure's stable bearing capacity is $P_{\alpha\ cr}$. The dimensionless quantity $\gamma$ defined by Equation represents the relative stable bearing capacity of the structure:

$$\gamma = \frac{P_{cr}^{\alpha}}{P_{cr}^{rigid}}. \qquad (6.13)$$

### 6.2.2 The Effect of Rotational Stiffness on Stable Bearing Capacity

As shown in Fig. 6.3 of the K6 single-layer shell structure, the span is 22 m and the pitch is 6 m, with fixed supports at the perimeter. The loading mode is the control load mode for spherical shell structures—uniform load over the entire span. From the perspective of structural stability and bearing capacity, the quantitative study is conducted on the relationship between $P_{cr}$ and $gra\_r_{mi}$ with the joint stiffness $\alpha$.

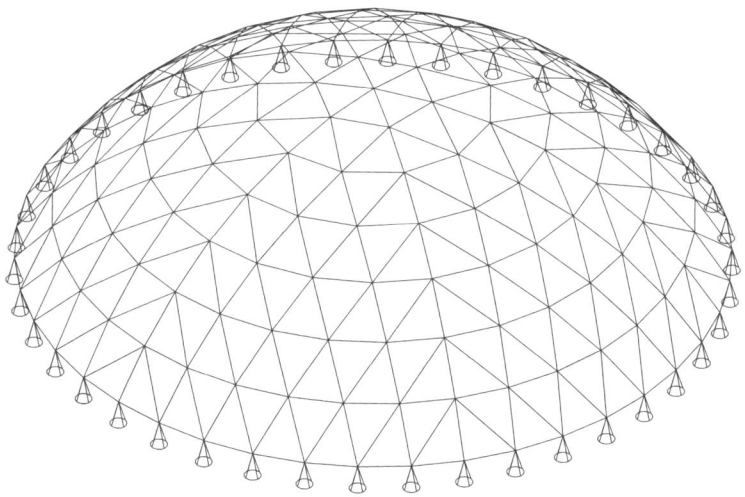

**Fig. 6.3** 22 m span single-layer spherical gridshell with a height of 6 m

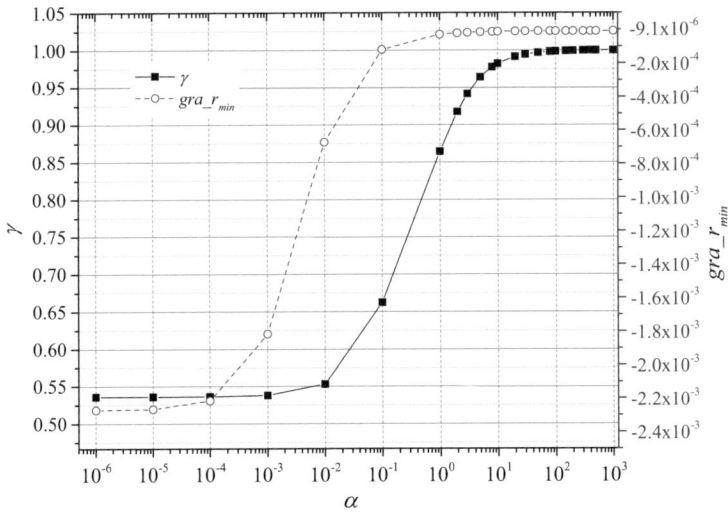

**Fig. 6.4** The relationship between γ and gra_$r_{min}$ as a function of α

Given the joint stiffness coefficient α, the stable carrying capacity $P_{cr}$ of the structure can be stably tracked by the arc length method. Then, the relative stable carrying capacity γ is obtained according to Equation. At the same time, based on Sect. 6.1, the corresponding structure's gra_rmin is calculated from the angle of joint stiffness degradation. By changing the joint stiffness coefficient α, the relationship between γ and gra_$r_{min}$, which represent the stability of the gridshell, can be obtained, as shown in Fig. 6.4.

From Fig. 6.4, the following 5 conclusions can be drawn:

(1) The joint stiffness has a significant impact on the stable bearing capacity of the gridshell. When the joint is an ideal hinge, its stable bearing capacity is about half lower than that of the gridshell with an ideal rigid joint.
(2) When α ∈ [$10^{-2}$, 10], the γ-α curve indicates that the structural bearing capacity under stable conditions significantly increases with the improvement of joint stiffness.
(3) When α ∈ [$10^{-2}$, 10], the gra_rmin-α curve indicates that the degree of structural stiffness degradation decreases significantly as the joint stiffness is increased.
(4) When α is greater than 10 or less than $10^{-2}$, the change in joint stiffness has little effect on the structural stability performance.
(5) Curve γ-α reflects the change in structural stable bearing capacity with the change of joint stiffness, while curve gra_rmin -α reflects the change in the trend of structural instability with the change of joint stiffness. The trend of γ and gra_rmin with respect to α is synchronized, which indicates that gra_rmin of joint stiffness can accurately reflect the influence of joint stiffness on the trend of structural instability from the perspective of stiffness degradation.

## 6.2.3 Range of Acceptable Joint Rotational Stiffness

Combining 6.2.2, when the joint is too large, increasing the joint stiffness cannot effectively improve the structural stability, but instead increases the steel usage at the joint. When the joint stiffness coefficient $\alpha$ is within the range of $[10^{-2}, 10^{1}]$, increasing the joint stiffness can significantly improve the structural stability. At the same time, in order to ensure that the structure has a reliable stable bearing capacity, it is stipulated that the joint stiffness coefficient $\alpha$ cannot be less than 1. Therefore, the appropriate range of joint stiffness coefficient $\alpha$ suitable for design is [1, 10].

The maximum number of bars that a joint can connect, as shown by the blue bars in Fig. 6.5, is called the main steel pipe of the joint. The corresponding outer diameter of the main steel pipe is recorded as the joint's main steel pipe outer diameter $P$. The joint's main steel pipe outer diameter $P$ is an important geometric parameter of the joint and also a geometric dimension for determining the joint's optimization parameters. By conducting topological optimization design on the joints with different sizes of main steel pipes within the reasonable range of joint stiffness, an optimized joint library can be obtained, as shown in Table 6.1.

For example, the joint with $P = 89$ mm can connect the maximum outer diameter of the member as 89 mm. Based on the maximum outer diameter $P$, the common member length of the main steel pipe can be determined according to experience, slenderness ratio, etc. The flexural stiffness of the member can be determined by the member length. When $\alpha$ is continuously taken in the range of [1, 10], the topology optimization design of the joint with fixed stiffness is carried out. The optimization model, technical route, and details of construction can be referred to Chap. 5. After the topology optimization design, the corresponding joint and joint mass are obtained. The maximum outer diameter of the member that these joints can connect is all 89 mm, but the joint stiffness and corresponding mass are different. The maximum joint stiffness is recorded as $k_{max}$, and the minimum joint stiffness is recorded as $k_{min}$.

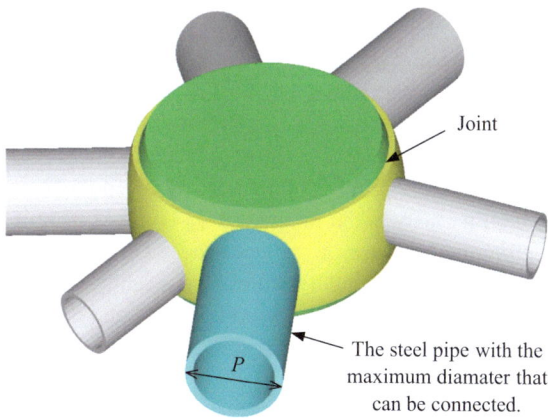

**Fig. 6.5** Joint and the maximum steel pipe it can connect

**Table 6.1** Optimized joints databese

| Joint | | Joint quality/kg | | | | | | | | | |
|---|---|---|---|---|---|---|---|---|---|---|---|
| Outside diameter of main steel pipe/mm | Pole length/m | α = 1 | α = 2 | α = 3 | α = 4 | α = 5 | α = 6 | α = 7 | α = 8 | α = 9 | α = 10 |
| Φ89 | 2.8 | 4.10 | 4.89 | 5.29 | 5.67 | 6.05 | 6.42 | 6.77 | 7.12 | 7.42 | 7.71 |
|  | 2.4 | 4.24 | 5.05 | 5.48 | 5.92 | 6.36 | 6.77 | 7.17 | 7.52 | 7.86 | 8.20 |
| Φ114 | 3.6 | 6.91 | 7.47 | 8.02 | 8.57 | 9.13 | 9.79 | 10.46 | 11.09 | 11.62 | 12.15 |
|  | 3.0 | 7.02 | 7.69 | 8.35 | 9.02 | 9.79 | 10.60 | 11.30 | 11.94 | 12.43 | 12.74 |
| Φ168 | 5.4 | 8.83 | 9.89 | 10.25 | 10.43 | 10.84 | 11.43 | 12.01 | 12.54 | 13.04 | 13.28 |
|  | 4.8 | 8.97 | 10.13 | 10.32 | 10.54 | 11.21 | 11.87 | 12.47 | 13.04 | 13.31 | 13.58 |

As stated in Sect. 5.2.1, the new joints have accompanying structural designs that are highly adaptable to the dimensions of the members. When the outer diameter of the steel tube D is less than or equal to the outer diameter of the main steel tube P in the joint, these steel tubes can all be connected to the joint, as shown by the gray members in Fig. 6.5. To avoid having too many types of joints, when the outer diameter of the main steel tube P is similar, larger joints can be used to replace smaller joints. For example, the joint corresponding to the outer diameter of the main steel tube $P = 114$ mm can connect the largest size member with a diameter of Φ114, and it can also connect the member with a diameter of Φ102. Since the line stiffness of the Φ114 member is similar to that of the Φ102 member, a separate joint is no longer needed for the outer diameter of the main steel tube $P = 102$ mm. Instead, the joint corresponding to the outer diameter of the main steel tube $P = 114$ mm is used to connect the Φ102 member. Similarly, the joint corresponding to the outer diameter of the main steel tube $P = 140$ mm is no longer needed and is replaced by the joint corresponding to the outer diameter of the main steel tube $P = 168$ mm.

Despite the fact that the values of dimensionless nodal stiffness α in Table 6.1 are discrete and the number of optimized joints is finite, the joint research results show that for the same type of joints with the same outer diameter of the main steel pipe P, the nodal stiffness is approximately linearly related to the nodal mass; at the same time, in the joint optimization technology, it is very convenient to obtain the corresponding optimized joint for any given α. Therefore, the nodal stiffness k can be continuously taken between $k_{min}$ and $k_{max}$. Thus, for the same type of joints, a continuous database of nodal stiffness and nodal mass can be established, and the database of partial nodal stiffness-mass relationships is shown in Fig. 6.6. Figure 6.6 also clearly indicates the $k_{min}$ and $k_{max}$ of a certain type of joint.

**Fig. 6.6** The relationship between joint stiffness and mass

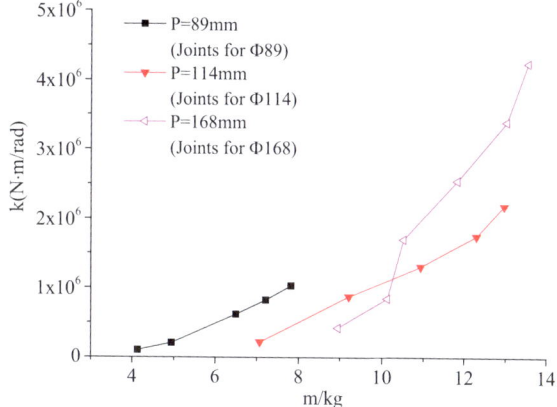

## 6.3 Title Stable Optimization of Single-Layer Gridshells with Consideration of Joint Stiffness

### 6.3.1 Optimization Goals

The instability mechanism of the gridshell indicates that the smaller $gra\_r_{min}$ is, the more significant the instability trend of the structure is, and the lower the stable bearing capacity $P_{cr}$ is. Compared with $P_{cr}$, $gra\_r_{min}$ is easier to calculate and does not require nonlinear iteration to express the stability of the structure. Therefore, the optimization goal is to maximize $gra\_r_{min}$, that is,

$$\text{Max } gra\_r_{min} \quad (6.14)$$

In the equation: $gra\_r_{min}$ is the minimum value of the gradient of the shape degree variation for all joint configurations, i.e.

$$gra\_r_{min} = \min(gra\_r_1, gra\_r_2 \ldots gra\_r_k \ldots gra\_r_n) \quad (6.15)$$

In the equation: $gra\_r_k$ represents the relative change gradient of the well-formedness of the $k$th elastic joint, which is calculated according to Equations; n is the total number of unconstrained joints.

### 6.3.2 Optimizing Variables

The beam and joint are both optimization variables. The beam variable is the index of the cross-section in the candidate cross-section list, which is a discrete variable. Based on the cross-section index and the candidate cross-section list, the outer diameter and wall thickness of the beam are determined. The joint variable is the outer diameter of the main steel pipe at the joint and the joint stiffness.

For the beam variable, the candidate section list is taken from the Chinese manufacturing standard GB/T 17395-2008 Dimensions, Shape, Weight and Permissible Deviations of Seamless Steel Tubes [3]. The beam variables are as follows:

$$\mathbf{I} = [\mathbf{I}_1, \mathbf{I}_2, \ldots, \mathbf{I}_k, \ldots, \mathbf{I}_{nm}] \tag{6.16}$$

In the equation: $\mathbf{I}_k$ represents the index of the $k$th beam section in the candidate section list; $nm$ is the number of beams; $\mathbf{I}$ is the optimization variable of the beam, representing the index of all beam sections in the candidate list, which is a vector of integer type.

Based on the optimized variable $\mathbf{I}$, the candidate list, and the information matrix $\mathbf{S}$ for the beam section dimensions can be determined.

$$\mathbf{S} = [\mathbf{S}_1, \mathbf{S}_2, \ldots, \mathbf{S}_k, \ldots, \mathbf{S}_{nm}]^T$$
$$\mathbf{S}_k = [D_k, t_k] \tag{6.17}$$

In the equation: $\mathbf{S}_k$ represents the cross-section size matrix of the $k$th member; $D_k$ and $t_k$ represent the outer diameter and wall thickness of the $k$th member respectively; $nm$ represents the number of members.

When determining the joint, the outer diameter of all members intersecting at the joint should be examined first, and the maximum outer diameter, i.e. the outer diameter of the main steel pipe P, should be determined, as shown in Fig. 6.5. From the outer diameter of the main steel pipe $P$, the appropriate joint type and the maximum value $k_{max}$ and minimum value $k_{min}$ of the joint stiffness corresponding to the type can be determined from Table 6.1 and Fig. 6.6. Therefore, the joint variable is described as follows.

$$\mathbf{J} = [\mathbf{J}_1, \mathbf{J}_2, \ldots, \mathbf{J}_k, \ldots, \mathbf{J}_n]^T$$
$$\mathbf{J}_k = [P_k, k_k] \tag{6.18}$$

In the equation: $\mathbf{J}$ is the matrix of joint variables; $P_k$ is the outer diameter of the main steel pipe at joint $k$; $k_k$ is the corresponding rotational stiffness, $k_k \in [k_{min}, k_{max}]$; $n$ is the total number of unconstrained joints.

### 6.3.3 Constraints

The steel usage constraints for members and joints are given by Equations and Equations, respectively. According to the GB 50017-2003 Steel Structure Design Code [4] and the JGJ 7-2010 Technical Code for Space Grid Structures [5], the design constraints for members are given by Equations and Equations.

(1) Constraint on steel usage for structural members:

$$V_{m,i} \leq V_{m,0} \quad (6.19)$$

In the equation, $V_{m,I}$ represents the steel usage per volume of the member after $i$ steps of optimization, and $V_{m,0}$ is the upper limit of the steel usage per volume of the member given.

(2) Steel volume constraint for joints:

$$m_{j,i} \leq m_{j,0} \quad (6.20)$$

In the equation, $m_{j,i}$ represents the steel consumption for joint quality after $I$ steps of optimization, and $m_{j,0}$ represents the upper limit of steel consumption for joint quality given.

(3) Strength constraint conditions:

$$\frac{N_i}{A_{ni}} \pm \frac{M_{xi}}{\gamma_x W_{nxi}} \pm \frac{M_{yi}}{\gamma_y W_{nyi}} \leq f \, (i = 1, 2, \ldots, nm) \quad (6.21)$$

In the equation: $N_i$ is the design value of the axial force of the $i$-th member; $A_{ni}$ is the net cross-sectional area of the $i$-th member; $M_{xi}$ and $M_{yi}$ are the design values of the bending moments in the two principal axes; $\gamma_x$ and $\gamma_y$ are the plastic development coefficients in the two directions; $W_{nxi}$ and $W_{nyi}$ are the net cross-sectional moduli in the two principal axes; $f$ is the design value of the material strength.

(4) Stability constraints for cantilever beams:

$$\frac{N_i}{\varphi_{yi} A_i} + \frac{\beta_{myi} M_{yi}}{\gamma_y W_{yi}(1 - 0.8 N_i / N_{Ei}')} + \eta \frac{\beta_{txi} M_{xi}}{\varphi_{bxi} W_{xi}} \leq f \, (i = 1, 2, \ldots, n_{cm}) \quad (6.22)$$

In the equation: $A_i$ is the gross cross-sectional area of the $i$-th member; $\varphi_{yi}$ is the stability coefficient of the axially compressed member; $\beta_{myi}$ and $\beta_{txi}$ are the equivalent moment coefficients in the plane and out of the plane, respectively; $W_{xi}$ and $W_{yi}$ are the gross cross-sectional moduli in the two principal axes; $\eta$ is the section influence coefficient; $\varphi_{bxi}$ is the overall stability coefficient of the uniformly curved flexural member; $N' Ei$ is the calculation parameter; $n_{cm}$ is the number of compression members.

## 6.3.4 Optimization Algorithm

In Sect. 3.2 of this paper, a stable optimization algorithm for gridshells based on the assumption of rigid joint connections, GGA, has been presented. After considering the joint stiffness, the specific steps of GGA are as follows:

(1) **Coding**: Binary coding is used, with the coding rules being identical to those in Sect. 3.2.2.
(2) **Joint configuration stiffness analysis and individual fitness calculation**: In accordance with Sect. 6.1, the stiffness of the structure is analyzed for each shell in the population with regard to its joint configuration, and the structure's $gra\_r_{min}$ is obtained. The constrained optimization model is transformed into an unconstrained problem by introducing a penalty function $p(x)$ through a penalty function method. The final individual fitness function is:

$$F(x) = p(x) \frac{1}{|gra\_r_{min}|} \qquad (6.23)$$

In this equation, $x$ represents a chromosome in the current population, and $p(x)$ is the penalty function, which is calculated according to Eq. (3.9) in Sect. 3.2.2.

(3) **Select superior individuals**: Sort all individuals in the population according to their individual fitness. Replicate the top quarter of the individuals twice, and the middle half once. This truncated selection is consistent with the one in Sect. 3.2.2.
(4) **Identify critical joints and critical members**: For weak joints and members, appropriate reinforcement should be provided to improve stability and bearing capacity. At the same time, redundant joints and members with low stiffness should be reduced. The identification of weak joints and members, as well as redundant joints and members, is as follows:

   (4.1) **Identify weak joints and weak members**: The $nvj$ lowest $gra\_r$ values among the joints tracked during stability analysis are generally the first to fail, defining them as the set of weak joints $\{j_v\}$. The weak joint $j_{v,k}$ ($k = 1, 2, \ldots nvj$) corresponds to the outer diameter of the main steel pipe $P_{v,k}$ and the bending stiffness $k_{v,k}$. Among all the members connected to the weak joint $j_{v,k}$ ($k = 1, 2, \ldots nvj$), the member with the smallest cross-sectional area is the weak member.

   (4.2) **Identify redundant joints and redundant members**: The top $nrj$ joints with the highest $gra\_r$ values in the stable tracking are generally stable, and they are defined as the redundant joint set $\{j_r\}$. The redundant joint $j_{r,k}$ ($k = 1, 2, \ldots nrj$) corresponds to the outer diameter of the main steel pipe $P_{r,k}$, and the bending stiffness is $k_{r,k}$. Among all the members connected to the redundant joint $j_{r,k}$ ($k = 1, 2, \ldots nrj$), the member with the largest cross-sectional area is the redundant member.

(5) **Guided mutate the critical members**: targeted adjustment of critical links to gradually approach the optimal solution. The specific mutation method is the same as Step 6 in Sect. 3.2.2.
(6) **Update the outer diameter of the pipes at critical joints**: In Step 5, the orientation of the critical members was varied, which may cause the outer diameter of the main steel pipe at related joints to change. The outer diameter of the main

6.3 Title Stable Optimization of Single-Layer Gridshells ...                149

steel pipe at the new joint must be updated, and the stiffness value closest to the original joint stiffness in the new joint stiffness range should be assigned to the new joint. The specific steps are as follows:

(6.1) For the weak joint $j_{v,k}$ ($k = 1, 2, \ldots nvj$), the updated outer diameter of the main steel pipe $P'_{v,k}$ and the updated joint stiffness $k'_{v,k}$ are calculated according to Eq. (6.24).

$$P'_{v,k} = P^{new}_{v,k}$$

$$k'_{v,k} = \begin{cases} k_{v,k}, & P_{v,k} = P^{new}_{v,k} \\ k^{new}_{max}, & k_{v,k} > k^{new}_{max} \& P_{v,k} \neq P^{new}_{v,k} \\ k_{v,k}, & k^{new}_{min} \leq k_{v,k} \leq k^{new}_{max} \& P_{v,k} \neq P^{new}_{v,k} \\ k^{new}_{min}, & k_{v,k} < k^{new}_{min} \& P_{v,k} \neq P^{new}_{v,k} \end{cases} \quad (6.24)$$

In the equation, $P^{new}_{v,k}$ is the new outer diameter of the main steel pipe at joint $j_{v,k}$ after the section adjustment in Step 5; $P_{v,k}$ and $k_{v,k}$ are from Step 4; $k^{new}_{max}$ and $k^{new}_{min}$ are the upper and lower limits of the joint's rotational stiffness corresponding to $P^{new}_{v,k}$.

(6.2) For redundant joint $j_{r,k}$ ($k = 1, 2, \ldots nrj$), the updated main pipe outer diameter $P'_{v,k}$ and the updated joint stiffness $k'_{r,k}$ are calculated according to Eq. (6.25).

$$P'_{r,k} = P^{new}_{r,k}$$

$$k'_{r,k} = \begin{cases} k_{r,k}, & P_{r,k} = P^{new}_{r,k} \\ k^{new}_{max}, & k_{r,k} > k^{new}_{max} \& P_{r,k} \neq P^{new}_{r,k} \\ k_{r,k}, & k^{new}_{min} \leq k_{r,k} \leq k^{new}_{max} \& P_{r,k} \neq P^{new}_{r,k} \\ k^{new}_{min}, & k_{r,k} < k^{new}_{min} \& P_{r,k} \neq P^{new}_{r,k} \end{cases} \quad (6.25)$$

In the equation, $P^{new}_{r,k}$ is the outer diameter of the main steel pipe at joint $j_{r,k}$ after the 5th step of section adjustment; $P_{r,k}$ and $k_{r,k}$ are from the 4th step; $k^{new}_{max}$ and $k^{new}_{min}$ are the upper and lower limits of the joint's rotational stiffness corresponding to $P^{new}_{r,k}$.

(7) **Guided mutatie critical joints**: For weak joints, appropriate strengthening of joint stiffness is needed; for redundant joints, appropriate reduction of joint stiffness is required, so that the population gradually approaches the optimal solution. The specific method is as follows:

(7.1) **Strengthen the weak Joints**: For the weak variant joint $j_{v,k}$ ($k = 1, 2, \ldots nvj$), the new joint stiffness knew v, k is given by Eq. (6.26).

$$k^{new}_{v,k} = \min\{k_{max}, (1 + ei)k_{v,k}'\} \quad (6.26)$$

In equation, $ei$ is the evolution rate of joint stiffness enhancement, which is recommended to be in the range of 4–6% regardless of the scale of the optimization object. Within this range, the optimization converges quickly. $k_{v,k}\prime$ is the current stiffness of joint $j_{v,k}$ after the main steel pipe diameter is updated in step 6.1, and $k_{\max}$ is the maximum stiffness of the current joint. Equation (6.26) shows that the new joint stiffness does not exceed $k_{\max}$.

(7.2) **Weaken the redundant joints**: For redundant joint $j_{r,k}$ ($k$=1, 2, … $nrj$), the new joint stiffness $k_{r,k}^{new}$ is given by Eq. (6.27).

$$k_{r,k}^{new} = \max\{k_{min}, (1-ed)k_{r,k}\prime\} \qquad (6.27)$$

In equation, $ed$ is the evolution rate of joint stiffness degradation, which is recommended to be set between 1 and 2% regardless of the scale of the optimization object. Within this range, the optimization converges quickly. $k_{r,k}\prime$ *is* the current stiffness of joint $j_{r,k}$ after the outer diameter of the main steel pipe is updated in step 6.2, and $k_{min}$ is the minimum stiffness of the current joint. Equation (6.27) shows that the new joint stiffness is no less than $k_{min}$.

(8) **Cutoff criterion**: If the algorithm evolves to a sufficient number of generations, stop optimizing; otherwise, proceed to step (2).

Similar to Sect. 3.2, the core of the GGA in this section is still the directed variation operation on the optimized variables, rather than the random variation operation in standard GA, thus improving the optimization efficiency; the difference from Sect. 3.2 lies in that this section's GGA needs to perform directed variation on both the beam element variables and joint variables. According to Sect. 6.1, identify the weak beam elements and joints, redundant beam elements and joints. First, perform directed variation on the weak beam elements to enhance these elements; perform directed variation on the redundant beam elements to weaken these elements. After the beam elements are enhanced, it may cause the maximum outer diameter of the intersecting beam elements at the joint to increase, making it impossible for the enlarged main steel pipe to be connected to the current joint; after the beam elements are weakened, it may cause the maximum outer diameter of the intersecting beam elements at the joint to decrease, making the current joint too stiff for the reduced main steel pipe. Therefore, after the beam element directed variation, the maximum outer diameter of intersecting beam elements at the joint $P$ should be recalculated, and the corresponding joint type should be determined in the joint library based on $P$. From the new joint type, select the joint with the closest stiffness to the original joint, and update the joint. After completing the joint update, appropriately enhance the joint stiffness of the weak joints within the limit of joint stiffness upper bound; for the excessively stiff joints, appropriately weaken them within the limit of joint stiffness lower bound. By orienting the member variations, the stiffness distribution of the members is optimized; by updating the joints, the appropriate size joints are connected to the members; by orienting the joint stiffness variations, the stiffness distribution of the joints is optimized. The specific process of GGA is shown in Fig. 6.7.

**Fig. 6.7** A guided genetic algorithm flow considering joint stiffness

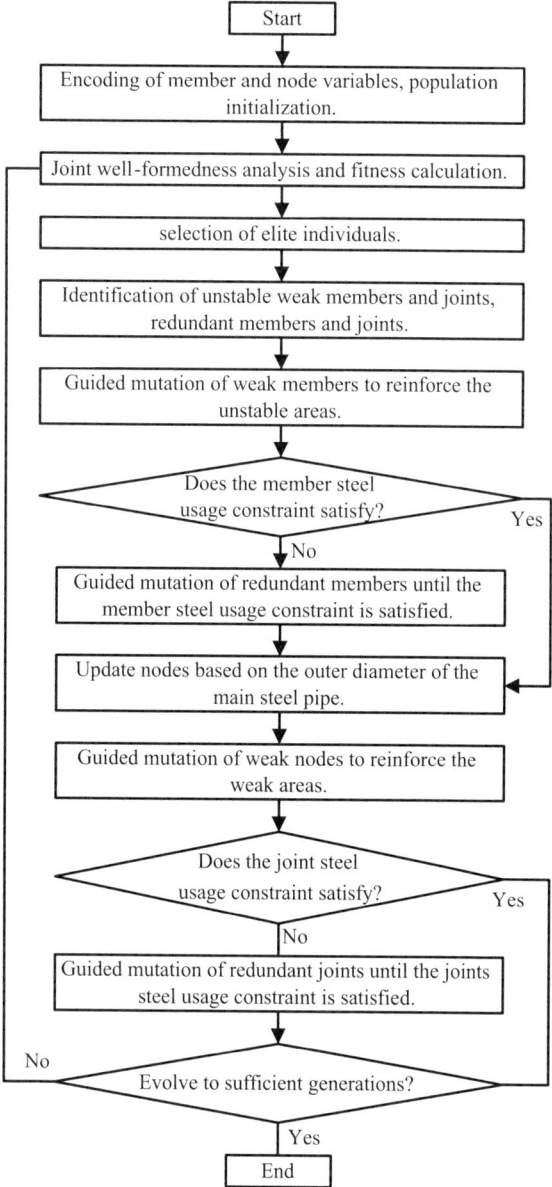

## 6.4 Optimization Example 1: Single-Layer Gridshell with 22 m Span

### 6.4.1 Stable Optimization Considering Joint Stiffness

In Sect. 3.3, a single-story K6 net shell with a span of 22 m and an eave height of 11 m is taken as an example and optimized for stability based on the assumption of rigid joints. The net shell is composed of 6 identical sectors, one of which is shown in bold in Fig. 6.8. It is made up of symmetrical red and black parts. The structure has a total of 462 members and 169 joints. After the stability optimization in Sect. 3.3, the optimized structure not only has a high stable bearing capacity ($P_{cr} = 94.22$ kN/m$^2$) but also saves steel usage (steel usage of 29.12 kg/m$^2$). According to the JGJ7-2010 Code for Design of Space Grid Structures, the joints of the optimized structure after Sect. 3.3 are designed. The joint form is a welded hollow ball joint, and the diameter and wall thickness of all hollow balls are taken to be the minimum value that meets the strength requirements. After the design, the steel usage of the joints is 3315 kg.

Using the stable optimization design method considering joint stiffness proposed in this chapter, the same single-layer shell structure in Sect. 3.3 is optimized for stability while determining the section and joints of the members. The relevant optimization parameters are as follows: the upper limit of steel used at the joints, $m_{j,0} = 2000$ kg; the evolution rate of joint stiffness enhancement in Equation, $ei = 5\%$, and

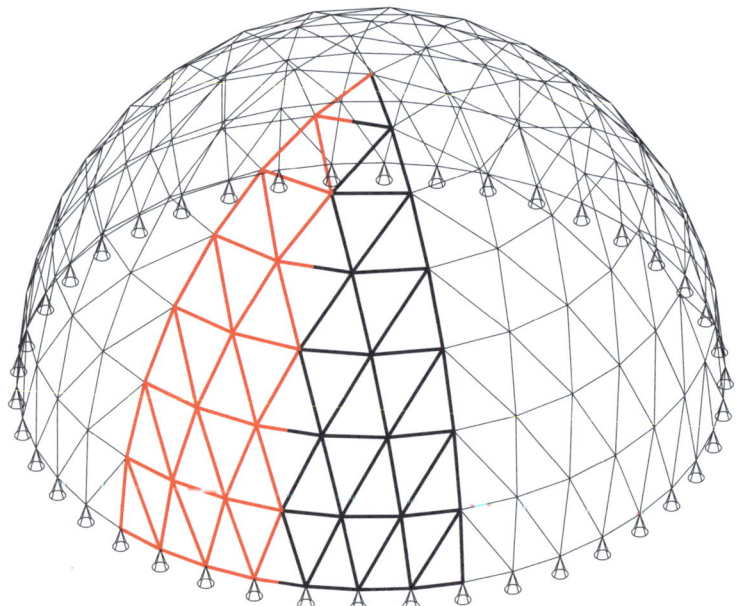

**Fig. 6.8** 22 m span single-layer spherical gridshell with a height of 11 m

## 6.4 Optimization Example 1: Single-Layer Gridshell with 22 m Span

the evolution rate of joint stiffness reduction in Equation, $ed = 1\%$; the steel used at the members and other optimization parameters are consistent with those in Sect. 3. 3.

The stable optimization algorithm considering nodal stiffness was run on a computer running Windows 7 operating system, which was configured with an Intel(R) Core(TM) i7-4790 K CPU@ 4.00 GHz and 32 GB of memory. At the 300th optimization step, the population evolved to the optimal solution. The time required to evolve to the 300th optimization step was 764.309 s, so the average time per generation was 2.548 s. The main steps of the stable optimization algorithm considering nodal stiffness and their corresponding computational times are shown in Table 6.2. 68% of the computational time in the optimization algorithm was spent on calculating the response of the netshell (i.e., the relative change in nodal well-formedness), which is basically consistent with the time distribution pattern of the genetic algorithm summarized by Adeli and Cheng [6].

Using the vertex and its surrounding 6 members shown in Fig. 6.8 as an example, the stable optimization design method considering joint stiffness is explained. After the 74th optimization step, the cross-sectional areas of the 6 members are all $\Phi 114 \times 4$ circular steel tubes, and the outer diameter of the main steel pipe at the joint is $P = 114$ mm. The maximum outer diameter of the member that can be connected to the joint is also 114 mm. The joint stiffness is $k = 2.19 \times 10^6$ N m/rad, as shown in Fig. 6.9a. In the 75th optimization step, first, according to Sect. 6.1, it can be calculated that the joint is significantly softened and is a weak joint, and the 6 members connected to it are weak members. After the 6 members are directed variation, the cross-sectional areas of all members are changed from $\Phi 114 \times 4$ to $\Phi 140 \times 3.2$. After the members are directed variation, the maximum outer diameter of the member's increases from 114 to 140 mm, and the original joint cannot connect the $\Phi 140$ member. At this time, the joint corresponding to $P = 168$ mm should be selected (the joint corresponding to $P = 140$ mm is merged into the joint corresponding to $P = 168$ mm, as described in Sect. 6.2.3). According to Fig. 6.6, the minimum and maximum values of k for this type of joint are $k_{min} = 4.25 \times 10^5$ N m/rad and $k_{max} = 4.25 \times 10^6$ N m/rad. In this type of joint, the joint with the same stiffness as the original joint (i.e., the joint corresponding to $P = 168$ mm and $k = 2.19 \times 10^6$ N m/rad) is assigned a value to the vertex, completing the joint update, as shown in Fig. 6.9b. After the update, the size of the joint can be guaranteed to connect with the member, and its stiffness is the same as the stiffness of the previous optimization step. Finally, the joint stiffness is directed variation according to Equation. A new joint stiffness $k = (1 + 5\%) \times 2.19 \times 10^6 = 2.30 \times 10^6$ N m/rad is obtained, which satisfies the condition of $k_{min} \leq k \leq k_{max}$. The value of $k = 2.30 \times 10^6$ N m/rad is assigned to the joint, as shown in Fig. 6.9c. After 4 steps of operation, namely identification-member orientation variation-joint update-joint orientation variation, the 75th optimization step is completed.

The optimization process of $gra\_r_{min}$ and $P_{cr}$ is shown in Fig. 6.10. The final optimized structure has $gra\_r_{min} = -2.252 \times 10^{-6}$, and a stable bearing capacity of $P_{cr} = 92.55$ kN/m². The steel usage per joint is 1988 kg $< m_{j,0}$, which meets the constraint requirement of steel usage per joint; the steel usage per member is 28.92 kg/

**Table 6.2** Optimization algorithm considering joint stiffness for a 22 m span gridshell: main steps and time

| Main steps | Corresponding functions | Function overview | Time/s | Percentage of total time consumed (%) |
|---|---|---|---|---|
| Population initialization | InitialGen | Generating initial members | 0.717 | 0.09 |
| | Initial_Joint_stiffness | Initial joint stiffness | 0.016 | 0.00 |
| Decoding and encoding | GenToPhenotype | Generating binary code for membrane | 91.823 | 12.01 |
| | GAencode | Generating binary code | 0.157 | 0.02 |
| | GAdecode | Translating binary code | 8.485 | 1.11 |
| Joint shape degree calculation and individual fitness calculation | KKEE | Integrating overall stiffness matrix | 110.746 | 14.49 |
| | Boundary | Handling boundary conditions | 11.009 | 1.44 |
| | StressMatrix | Integrating geometric stiffness matrix | 117.654 | 15.39 |
| | MechanicsConstrain | Calculating mechanical constraint terms and penalty function | 152.509 | 19.95 |
| | FitnessCalculation | Calculating deformation gradient and individual fitness | 131.103 | 17.16 |
| Oriented mutation of rods | Find_variable | Identifying variables corresponding to the variant members | 1.232 | 0.16 |
| Oriented mutation of joints | Joint_Diameter_renew | Updating the diameter of the joint after variant members are altered | 3.132 | 0.41 |
| | Joint_stiffness_renew | Orienting the variant joint stiffness | 5.977 | 0.78 |
| File reading and writing | WriteBest | Outputting optimization results | 60.112 | 7.86 |
| Other | – | – | 69.637 | 9.11 |
| Total running time | – | – | 764.309 | 100.00 |

## 6.4 Optimization Example 1: Single-Layer Gridshell with 22 m Span

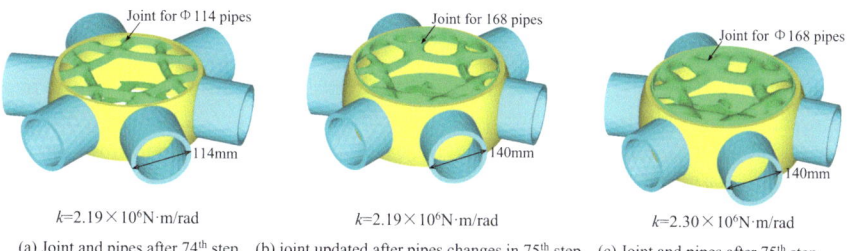

(a) Joint and pipes after 74th step  (b) joint updated after pipes changes in 75th step  (c) Joint and pipes after 75th step

**Fig. 6.9** Evolution of joints and connected members

m$^2$, and the total steel usage per member is 1.39 m$^3$ < $V_{m,0}$ = 1.41 m$^3$, which meets the constraint requirement of steel usage per member. Figure 6.10 shows that: considering the stiffness of the joints in the stable optimization algorithm can maximize $gra\_r_{min}$, reduce the softening degree of the structure, and significantly improve the structural stable bearing capacity; when the search is conducted near the optimal solution, $gra\_r_{min}$ and $P_{cr}$ remain stable, indicating that the stable optimization algorithm considering the stiffness of the joints can converge to the optimal solution.

After considering the stability optimization of joint stiffness, the beam section distribution is shown in the upper half of Fig. 6.11. When it is assumed that the joints are ideal rigidly connected, the beam distribution determined in Sect. 3.3 is shown in the lower half of Fig. 6.11. Since the structure is symmetrical and the load is symmetrical, the beam distribution is also symmetrical, so the 1/12 structure shown in red in Fig. 6.8 (see Fig. 6.11) is selected for explanation. In Fig. 6.11, the black beams have the same section in both optimization processes, while the red beams have different sections in the two optimization processes. Comparing the upper and lower parts of Fig. 6.11 shows that: (1) In the case where the optimization parameters are

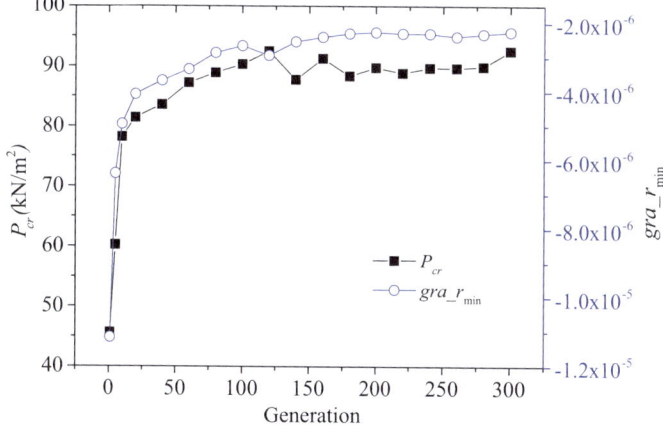

**Fig. 6.10** The optimization process of 22 m span mesh shell structure with $gra\_r_{min}$ and $P_{cr}$

the same, more than half of the black beams indicate that most of the beams have the same section size in the two optimization processes; (2) Some of the red beams have the same outer diameter in both optimization processes, with only the wall thickness being slightly different, such as Φ114 × 3.2 and Φ114 × 3.5; (3) The remaining red beams have an increased (or decreased) outer diameter in the second optimization process, while the wall thickness is correspondingly reduced (or increased), such as Φ140 × 3 and Φ114 × 4; (4) For all red beams, although the dimensions are slightly different in the two optimization processes, the beam stiffness is almost the same. The $P_{cr}$ of the optimized structure shown in the upper half of Fig. 6.11 is $P_{cr} = 92.55$ kN/m$^2$, which is only slightly smaller than the $P_{cr}$ of the structure shown in the lower half of Fig. 6.11 ($P_{cr} = 94.22$ kN/m$^2$). In summary, both from the level of local beam section dimensions and the level of overall structure stability and bearing capacity, it can be concluded that the stability optimization design method considering joint stiffness can obtain a reasonable beam distribution.

After the stable optimization of nodal stiffness, there are a total of 4 elastic joints in the structure. The location and numbering of the elastic joints are shown in the upper half of Fig. 6.11, and the joint characteristics are listed in Table 6.3. Since the constraint conditions are fixed supports on all sides, the joints at the supports are still traditional welded hollow ball joints. Combining the beam distribution shown in Fig. 6.11, it can be seen that the outer diameter of the main steel pipe at the elastic joints is no less than the maximum outer diameter of the intersecting steel pipes at the joint. This shows that the joint update operation in the optimization algorithm ensures that the joints have sufficient size to connect the adjacent members. At the same time, the outer diameter of the main steel pipe at the joint is equal to or slightly

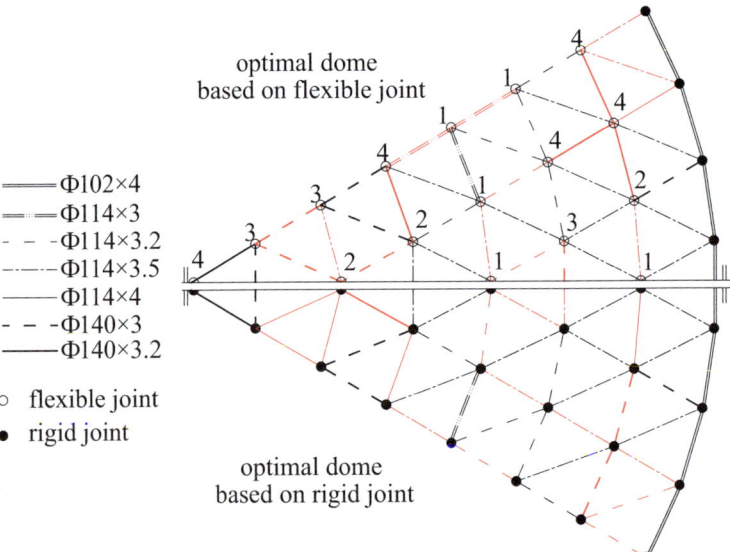

**Fig. 6.11** Sections and joints of two optimized domes

## 6.4 Optimization Example 1: Single-Layer Gridshell with 22 m Span

**Table 6.3** Joints of the optimized 22 m span single-layer gridshell

| Joint number | P/mm | m/kg | k/$10^6$ N m | Joint number | P/mm | m/kg | k/$10^6$ N m |
|---|---|---|---|---|---|---|---|
| 1 | 114 | 12.83 | 2.10 | 3 | 168 | 13.28 | 3.82 |
| 2 | 168 | 12.86 | 3.30 | 4 | 168 | 13.42 | 4.03 |

**Table 6.4** Optimized gridshells with 22 m span obtained by two optimization methods

|  | Optimized gridshell based on rigid joints | Optimized gridshell considering joint stiffness |
|---|---|---|
| $P_{cr}$ (kN/m$^2$) | 94.22 | 92.55 |
| gra_$r_{min}$ ($10^{-6}$) | − 1.789 | − 2.252 |
| Steel consumption for members (m$^3$) | 1.40 | 1.39 |
| Steel consumption for joints (kg) | 3315 | 1988 |

larger than the maximum outer diameter of the intersecting steel pipes at the joint, which shows that the optimization algorithm can avoid the joint size being too large and search for the most reasonable joint size.

The optimized results considering the joint stiffness are shown in Table 6.4. The optimized section distribution of the members based on the rigid joint optimization is shown in Sect. 3.3. The joint is also an ideal rigidly connected joint, so the maximum stable bearing capacity $P_{cr}$ of the corresponding structure is 94.22 kN/m$^2$. The single-layer shell structure obtained by considering the joint stiffness in the stable optimization design method has a member stiffness distribution similar to that of the structure obtained in Sect. 3.3, with $P_{cr}$ = 92.55 kN/m$^2$. Although it is slightly smaller than 94.22 kN/m$^2$, the steel usage at the joints is significantly reduced. This shows that the range of joint stiffness proposed in Sect. 6.2.3 is reasonable. The joint stiffness within this range can not only reduce the steel usage at the joints, but also ensure the stable bearing capacity of the structure.

### 6.4.2 The Anti-collapse Performance of Structures with Stable Optimized Stiffness After Joint Consideration

Based on the assumption of rigid joint, after stable optimization, the structural stable bearing capacity can be greatly improved. As mentioned in the previous section, when the joint is transformed from an ideal rigid joint to a semi-rigid joint, the stable bearing capacity of the semi-rigid joint gridshell is only 1.78% lower than that of the ideal rigidly connected joint gridshell. Therefore, after the joint is changed from rigidly connected to semi-rigidly connected, the stable bearing capacity of the optimized structure is still higher than the requirement of the code, but its ability to resist collapse under seismic action becomes a more concerned aspect for designers.

By comparing the anti-collapse performance of rigid joint truss structures and semi-rigid joint truss structures, the anti-collapse ability of the truss structure after considering the stability optimization of joint stiffness is verified. First, a numerical model based on the rigid joint assumption is established for the truss structure optimized for joint stiffness in Sect. 3.3. In the numerical model, the joints are assumed to be ideal rigidly connected, and the materials are ideal elasto-plastic. The representative value of the gravity load is 1.0 constant load + 0.5 live load, which is established as an equivalent joint concentrated mass at the joint using a mass element.

After completing the rigid joint shell model, a numerical model is established for the shell structure optimized for joint stiffness in Sect. 6.4.1. In the numerical model, the rotational stiffness $k$ of the semi-rigid joints is taken as specified in Table 6.3. By applying a plane-out-of-plane moment $M_y$ (as shown in Fig. 5.5) to the optimized joints, the maximum stress in the joint core under $M_y$ can be obtained. Since the joint is in the linear elastic state, the corresponding plane-out-of-plane yield moment $M_y^{yield}$ can be calculated when the maximum stress reaches the yield stress. For example, by applying a unit plane-out-of-plane moment (i.e., $M_y$) to each joint end of the optimized joint in Sect. 5.2.7, the stress distribution of the joint can be obtained, as shown in Fig. 5.13, with the maximum stress of $1.84 \times 10^4$ N/m². The yield stress of the material is $345 \times 10^6$ N/m², so when the joint end moment is $(345 \times 10^6)/(1.84 \times 10^4) \times 1 = 18.75$ kN m, the joint begins to enter the plastic state, and a conservative value of the joint's plane-out-of-plane yield moment $M_y^{yield}$ is assumed, assuming that the moment does not increase when the joint reaches $M_y^{yield}$. Similarly, the joint's plane-in-plane yield moment $M_z^{yield}$ can be obtained. Combining Figs. 5.5 and 5.6, the maximum stress of the optimized joint in the plane-out-of-plane moment $M_y$ and the plane-in-plane moment $M_z$ is basically the same; since the joint has the same rotational stiffness in the plane-in-plane and plane-out-of-plane directions, a conservative value of the joint's yield moment $M^{yield}$ is assumed to be the minimum of $M_y^{yield}$ and $M_z^{yield}$, as shown in Fig. 6.12. Perform stress analysis on the joints M-θ in Table 6.3 to obtain the yielding moment at each joint, as shown in Table 6.5. Because the rotational stiffness direction of the joint varies with the position and orientation of the connected members, a different coordinate system should be established at each semi-rigid connection point in the numerical model. The joint stiffness and yielding moment should be accurately assigned to the corresponding spring connection element in the model.

Modal analysis was conducted on two gridshells, and the first three natural frequencies of the structures are shown in Table 6.6. Table 6.6 indicates that: (1) the vibration frequencies of the semi-rigid joint gridshell are lower than those of the rigid joint gridshell, which is consistent with the concept of the structure; (2) the vibration frequencies of the semi-rigid joint gridshell are slightly lower than the corresponding frequencies of the rigid joint gridshell at the same order, which shows that the optimization algorithm selected joint stiffness has not reduced the dynamic performance of the structure in the linear elastic stage; (3) the frequency distribution

## 6.4 Optimization Example 1: Single-Layer Gridshell with 22 m Span

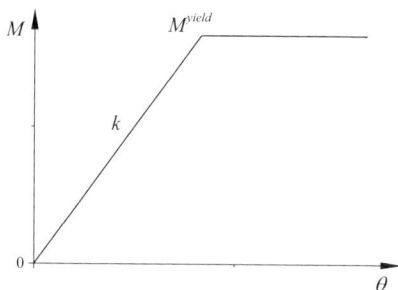

**Fig. 6.12** Semi-rigid joint M-θ

**Table 6.5** Mechanical properties of semi-rigid joints in anti-collapse analysis of 22 m span shell structure

| Joint number | P/mm | m/kg | k/$10^6$ N m | $M^{yield}$/kN m |
|---|---|---|---|---|
| 1 | 114 | 12.83 | 2.10 | 6.436 |
| 2 | 168 | 12.86 | 3.30 | 6.862 |
| 3 | 168 | 13.28 | 3.82 | 7.480 |
| 4 | 168 | 13.42 | 4.03 | 11.889 |

of the two structures is that the first and second order frequencies are equal, and the third order frequency is greater than the first and second order frequencies, which verifies the numerical model of the semi-rigid joint gridshell.

For the two verified numerical models, the El-centro earthquake wave is input to study the collapse performance of the structure. Since the netshell is a spatial structure and sensitive to vertical seismic motion, a three-dimensional seismic motion input is used, with the peak acceleration ratios of X, Y, and Z directions being 1:0.85:0.65, as shown in Fig. 6.13.

Increasing the peak ground acceleration (PGA) gradually, nonlinear time history analysis was conducted on two structures in the ANSYS platform, yielding load–displacement curves as shown in Fig. 6.14. The maximum PGA that the rigid joint truss structure could withstand was 4.0 g, and the load–displacement curve became increasingly flat as the PGA increased, indicating that the optimized truss structure obtained based on the rigid joint assumption exhibits nonlinear deformation of the structure with increasing PGA, showing a predictable collapse failure mode. The maximum PGA that the semi-rigid joint truss structure could withstand was 3.2 g. When PGA was less than 1 g, the maximum deformation of the semi-rigid joint truss structure was almost the same as that of the rigid joint truss structure, which was

**Table 6.6** First three frequencies of optimized gridshells with two 22 m spans

| | 1st frequency/Hz | 2nd frequency/Hz | 3rd frequency/Hz |
|---|---|---|---|
| Ideal rigidly connected gridshell | 5.2548 | 5.2548 | 8.1462 |
| Semi-rigidly connected gridshell | 5.1729 | 5.1729 | 7.9838 |

verified by the modal analysis results. As PGA increased, the deformation of the semi-rigid joint truss structure gradually exceeded that of the rigid joint truss structure, and the load–displacement curve was more flat than that of the rigid joint truss structure, indicating that the semi-rigid joint truss structure exhibits more obvious precursors of collapse than the rigid joint truss structure before collapse. The collapse critical load of the semi-rigid joint truss structure was 3.2 g, slightly less than the collapse critical load of the rigid joint truss structure (4.0 g), but far exceeded the peak acceleration of the rare earthquake with a return period of 900 years (0.62 g) stipulated in China's earthquake resistance code, and its collapse precursors were more obvious.

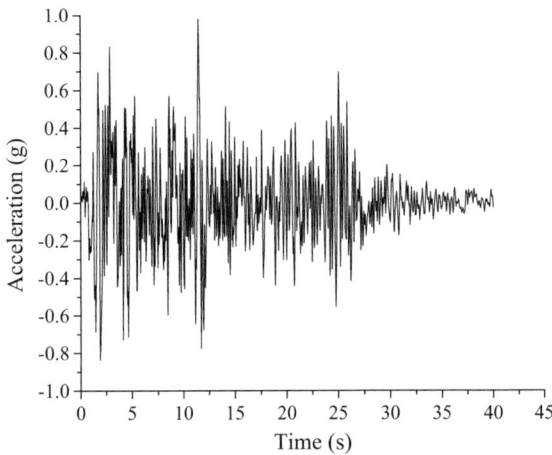

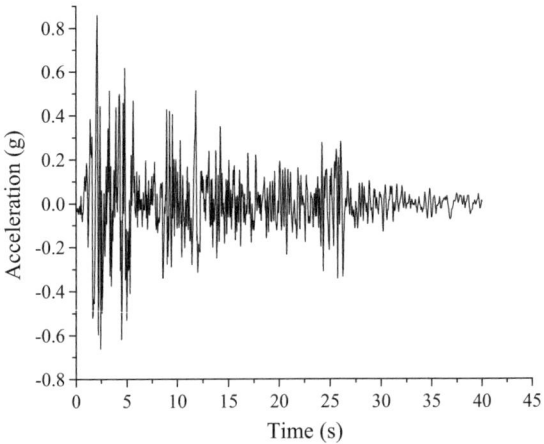

**Fig. 6.13** El-centro P-wave velocity time history

(a) Acceleration Time History in the X-Direction

(b) Acceleration Time History in the Y-Direction

**Fig. 6.13** (continued)

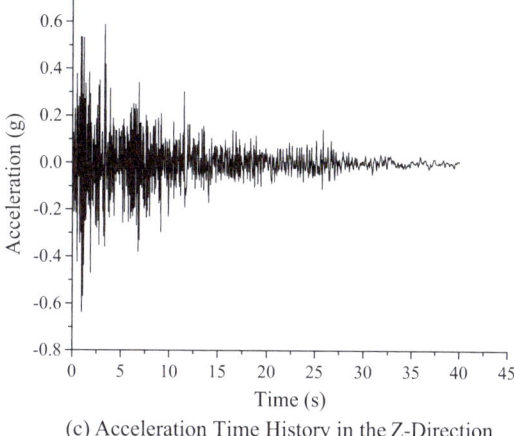

(c) Acceleration Time History in the Z-Direction

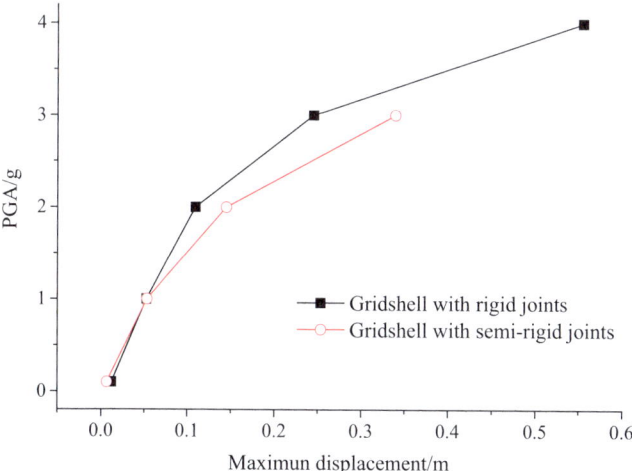

**Fig. 6.14** PGA-displacement curves for rigid-joint and semi-rigid-joint tensile gridshells with 22 m span

## 6.5 Optimization Example 2: Single-Layer Gridshell with 50 m Span

### 6.5.1 Stable Optimization Considering Joint Stiffness

In Sect. 3.4, a single-layer K6 netshell with a span of 50 m and a pitch of 20 m is taken as an example and optimized for stability based on the assumption of rigid joints. The netshell consists of six identical sectors, one of which is shown in bold in Fig. 6.15. It is composed of symmetrical red and black parts. The structure has a total of 1122

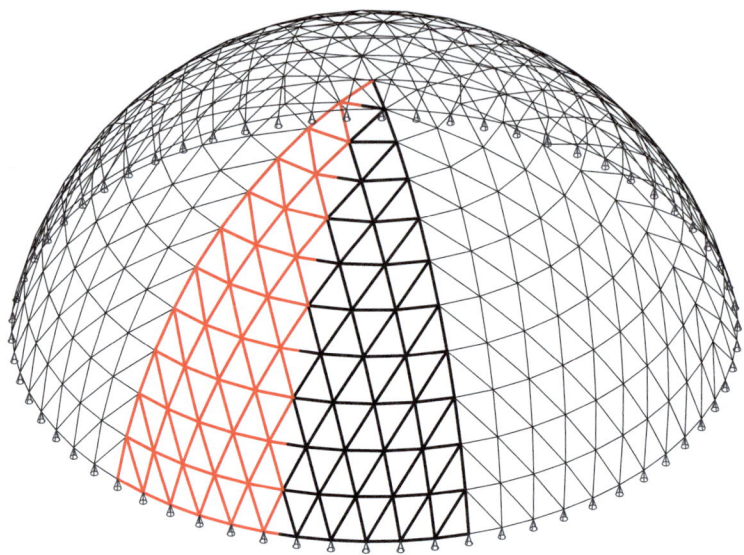

**Fig. 6.15** 50 m span single-layer spherical gridshell

members and 397 joints. After the stability optimization in Sect. 3.4, the optimized structure not only has a high stable bearing capacity ($P_{cr} = 32.55$ kN/m$^2$) but also saves steel usage (steel usage of 24.01 kg/m$^2$). According to the JGJ7-2010 Code for Design of Space Grid Structures, the joint design of the optimized structure in Sect. 3.4 is carried out. The joint form is a welded hollow ball joint, and the diameter and wall thickness of all hollow balls are taken to be the minimum value that meets the strength requirements. After the design, the steel usage of the joints is 16,440 kg.

Using the stable optimization design method considering joint stiffness proposed in this chapter, the same single-layer shell structure in Sect. 3.4 is optimized for stability while determining the section and joints of the members. The relevant optimization parameters are as follows: the upper limit of steel used at the joints, $m_{j,0} = 5500$ kg; the evolution rate of joint stiffness enhancement in the Equation, $ei = 5\%$, and the evolution rate of joint stiffness weakening in the Equation, $ei = 1\%$; the steel used at the members and other optimization parameters are consistent with those in Sect. 3.4.

The stable optimization algorithm considering joint stiffness runs on a computer operating on Windows 7, which is configured with an Inter(R) Core(TM) i7-4790 K CPU@ 4.00 GHz and 32 GB of memory. The population evolves to the optimal solution at the 200th optimization step. The time required to evolve to the 200th optimization step is 1988.353 s, so the average time per generation is 9.942 s. The main steps of the stable optimization algorithm considering joint stiffness and their corresponding times are shown in Table 6.7. When the structure size is expanded from 22 m span to 50 m span, the time spent computing the structure response increases from 68 to 84% of the total time.

## 6.5 Optimization Example 2: Single-Layer Gridshell with 50 m Span

**Table 6.7** Optimization algorithm considering joint stiffness for a 50 m span gridshell: main steps and time

| Main steps | Corresponding functions | Function overview | Time/s | Percentage of total time (%) |
|---|---|---|---|---|
| Population initialization | InitialGen | Generating initial members | 0.421 | 0.02 |
| | Initial_Joint_stiffness | Initial joint stiffness | 0.031 | 0.00 |
| Decoding and encoding | GenToPhenotype | Generating binary code for membrane | 111.725 | 5.62 |
| | GAencode | Generating binary code | 0.092 | 0.00 |
| | GAdecode | Translating binary code | 9.484 | 0.48 |
| Joint shape degree calculation and individual fitness calculation | KKEE | Integrating overall stiffness matrix | 256.831 | 12.92 |
| | Boundary | Handling boundary conditions | 56.572 | 2.85 |
| | StressMatrix | Integrating geometric stiffness matrix | 289.436 | 14.56 |
| | MechanicsConstrain | Calculating mechanical constraint terms and penalty function | 211.743 | 10.65 |
| | FitnessCalculation | Calculating deformation gradient and individual fitness | 853.494 | 42.92 |
| Oriented mutation of rods | Find_variable | Identifying variables corresponding to the variant members | 1.206 | 0.06 |
| Oriented mutation of joints | Joint_Diameter_renew | Updating the diameter of the joint after variant members are altered | 5.201 | 0.26 |
| | Joint_stiffness_renew | Orienting the variant joint stiffness | 9.857 | 0.50 |
| File reading and writing | WriteBest | Outputting optimization results | 60.26 | 3.03 |

(continued)

Table 6.7 (continued)

| Main steps | Corresponding functions | Function overview | Time/s | Percentage of total time (%) |
|---|---|---|---|---|
| Other | – | – | 122 | 6.14 |
| Total running time | – | – | 1988.353 | 100.00 |

The optimization process of $P_{cr}$ and $gra\_r_{min}$ is shown in Fig. 6.16. The final optimized structure has $P_{cr} = 32.11$ kN/m$^2$ and $gra\_r_{min} = -3.887 \times 10^{-6}$. The steel usage per joint is 55471 kg $< m_{j,0}$, which meets the constraint requirement of steel usage per joint. Figure 6.16 shows that for this medium-span single-layer shell structure, the optimization design method can achieve the maximum $gra\_r_{min}$, significantly reducing the softening degree of the structure and improving its stable bearing capacity.

After considering the stable optimization of joint stiffness, the section distribution of the members is shown in Fig. 6.17. The elastic joint positions and numbers are shown in Fig. 6.17, and the specific characteristics of the elastic joints are shown in Table 6.8. Because the structure is symmetrical and the load is symmetrical, the distribution of the stiffness of the members and joints optimized by the design method is also symmetrical. Therefore, the 1/12 structure (the red part in Fig. 6.15) is selected for explanation from Fig. 6.17. There are a total of 6 types of sections and 11 types of elastic joints in the optimized structure. Because the constraint condition is a fixed support on the perimeter, the joints at the support still use the traditional welded hollow ball joint.

The stable optimization results considering the joint stiffness are compared with those of Sect. 3.4 in Table 6.9. In Sect. 3.4, the optimal section distribution of the

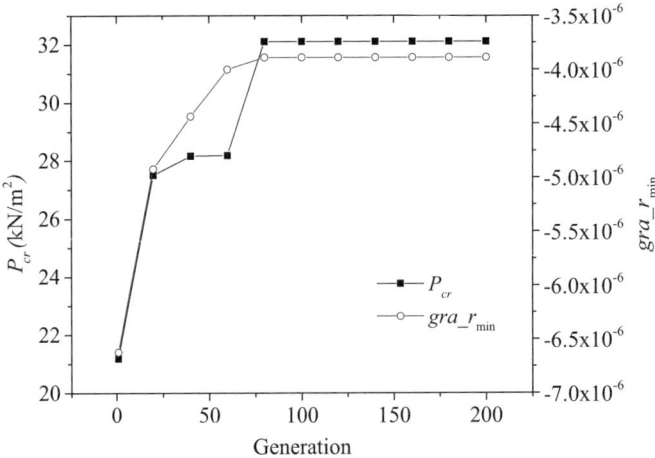

**Fig. 6.16** The optimization process of 50 m span mesh shell structure with $gra\_r_{min}$ and $P_{cr}$

## 6.5 Optimization Example 2: Single-Layer Gridshell with 50 m Span

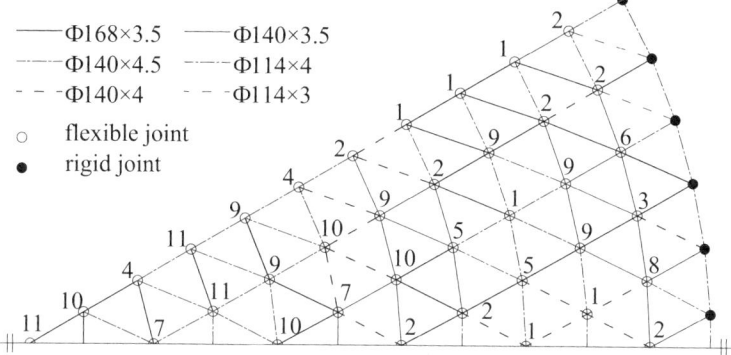

**Fig. 6.17** Sections and joints of an optimized 50 m span single-layer gridshell

**Table 6.8** Joints of the optimized 50 m span single-layer gridshell

| Joint number | P/mm | m/kg | k/$10^6$ N m | Joint number | P/mm | m/kg | k/$10^6$ N m |
|---|---|---|---|---|---|---|---|
| 1 | 168 | 12.00 | 2.67 | 7 | 168 | 14.72 | 6.13 |
| 2 | 168 | 13.10 | 3.53 | 8 | 168 | 14.79 | 6.23 |
| 3 | 168 | 13.69 | 4.47 | 9 | 168 | 14.95 | 6.50 |
| 4 | 168 | 14.31 | 5.47 | 10 | 168 | 15.10 | 6.73 |
| 5 | 168 | 14.43 | 5.67 | 11 | 168 | 15.18 | 8.67 |
| 6 | 168 | 14.64 | 6.00 | | | | |

members and ideal rigidly connected joints were obtained through optimization, so the $P_{cr}$ of the corresponding structure was the highest, with a value of 32.55 kN/m². However, the steel used in the joints was as high as 16,440 kg. By using the stable optimization design method of this chapter, which considers the influence of joint stiffness and member stiffness on the stability performance of the structure and optimizes the joint, the final $P_{cr}$ value of the structure is 32.11 kN/m², slightly smaller than 32.55 kN/m², but the steel used in the joints is only 33.3% of the steel used by the conventional design method. This further verifies the reasonable range of joint stiffness and shows that the joints developed in Chap. 5 are efficient, with a reasonable distribution of joint stiffness.

### 6.5.2 The Anti-collapse Performance of Structures with Stable Optimized Stiffness After Joint Consideration

By comparing the collapse resistance performance of rigid joint truss structures and semi-rigid joint truss structures, the collapse resistance ability of the truss structure

**Table 6.9** Optimized gridshells with 50 m span obtained by two optimization methods

|  | Optimized gridshell based on rigid joints | Optimized gridshell considering joint stiffness |
|---|---|---|
| $P_{cr}$ (kN/m$^2$) | 32.55 | 32.11 |
| $gra\_r_{min}$ ($10^{-6}$) | −3.576 | −3.887 |
| Steel consumption for members (m$^3$) | 6.04 | 6.04 |
| Steel consumption for joints (kg) | 16,440 | 5471 |

optimized for stability considering joint stiffness in Sect. 6.5.1 is verified. First, a numerical model based on the rigid joint assumption is established for the truss structure optimized in Sect. 3.4. In the numerical model, it is assumed that the joints are ideal rigidly connected, and other modeling methods are the same as those in Sect. 6.4.2. After the rigid joint truss model is established, the truss structure optimized for stability considering joint stiffness in Sect. 6.5.1 is established numerically. The yield moment of the optimized joints in Table 6.8 is determined according to the method in Sect. 6.4.2, so the joint stiffness and yield moment of the truss structure after stability optimization are shown in Table 6.10. When establishing semi-rigid joint connections, a coordinate system is established at each semi-rigid connection point, and the stiffness and yield moment of the corresponding spring connection unit are accurately assigned.

Modal analysis was conducted on two cable net structures, and the first three natural frequencies of the structures are shown in Table 6.11. Table 6.11 indicates that: (1) The vibration frequencies of the semi-rigid joint cable net structure are lower than those of the rigid joint cable net structure, which is consistent with the concept of

**Table 6.10** Mechanical properties of semi-rigid joints in anti-collapse analysis of 50 m span shell structure

| Joint number | P/mm | m/kg | k/$10^6$ N m | $M^{yield}$/kN m |
|---|---|---|---|---|
| 1 | 168 | 12.00 | 2.67 | 5.496 |
| 2 | 168 | 13.10 | 3.53 | 7.196 |
| 3 | 168 | 13.69 | 4.47 | 7.890 |
| 4 | 168 | 14.31 | 5.47 | 8.403 |
| 5 | 168 | 14.43 | 5.67 | 8.503 |
| 6 | 168 | 14.64 | 6.00 | 8.677 |
| 7 | 168 | 14.72 | 6.13 | 8.743 |
| 8 | 168 | 14.79 | 6.23 | 8.801 |
| 9 | 168 | 14.95 | 6.50 | 8.933 |
| 10 | 168 | 15.10 | 6.73 | 9.058 |
| 11 | 168 | 15.18 | 8.67 | 9.124 |

the structure; (2) The vibration frequencies of the semi-rigid joint cable net structure are slightly lower than the corresponding frequencies of the rigid joint cable net structure at the same order, indicating that the optimization algorithm selected the joints without significantly reducing the dynamic performance of the structure in the linear elastic stage; (3) The frequency distribution patterns of the two structures are the same, verifying the numerical model of the semi-rigid joint cable net structure.

The same three-dimensional El-centro wave input as in Sect. 6.4.2 was used to perform nonlinear time history analyses of the two structures, step by step, with increasing seismic ground motion PGA. The load–displacement curves for the rigid-joint truss structure and the semi-rigid-joint truss structure are shown in Fig. 6.18. The maximum PGA that the rigid-joint truss structure can withstand is 5.0 g, and the load–displacement curve becomes increasingly flat as the PGA is increased, indicating that the optimized truss structure based on the rigid-joint assumption experiences nonlinear deformation with increasing PGA and exhibits a predictable collapse failure mode under seismic action. The maximum PGA that the semi-rigid-joint truss structure can withstand is 1.6 g. When PGA is less than 1.0 g, the maximum deformation of the semi-rigid-joint truss structure is almost the same as that of the rigid-joint truss structure, which is verified by the modal analysis results. When PGA is greater than 1.0 g, the curvature of the load–displacement curve of the semi-rigid-joint truss structure first becomes increasingly flat. The collapse critical load of the semi-rigid-joint truss structure is 1.6 g, which is smaller than the collapse critical load of the rigid-joint truss structure (5.0 g). From the perspective of plastic hinge development, the reasons for the reduction of the collapse critical load of the semi-rigid-joint truss structure are as follows: (1) The ratio of the steel used in the joints to the steel used in the members of the truss structure in Sect. 6.4 is 0.18:1. In this section, the steel usage ratio of the joint is 0.11:1, which is significantly smaller than that of the truss structure in Sect. 6.4, resulting in some joints yielding before the members do (the article considers that once the maximum stress at the joint reaches the yield stress, the entire joint is considered to have yielded). This forms a plastic hinge at the joint, leading to local structural instability; (2) The spherical curvature radius of the truss in this section is 25.625 m, while the curvature radius of the truss structure in Sect. 6.4 is 11 m, so the truss surface in this section is flatter. After the plastic hinge forms, it is more likely to form structural instability under seismic cyclic loading. Although the collapse critical load of the semi-rigid joint truss structure is less than that of the rigid joint truss structure, the collapse critical load is already far beyond the peak acceleration value (0.62 g) of the 9° rare earthquake, and its collapse precursor is more obvious.

**Table 6.11** First three frequencies of optimized gridshells with two 50 m spans

|  | 1st frequency/Hz | 2nd frequency/Hz | 3rd frequency/Hz |
| --- | --- | --- | --- |
| Ideal rigidly connected gridshell | 4.6095 | 4.6095 | 6.1709 |
| Semi-rigidly connected gridshell | 4.5565 | 4.5565 | 6.0689 |

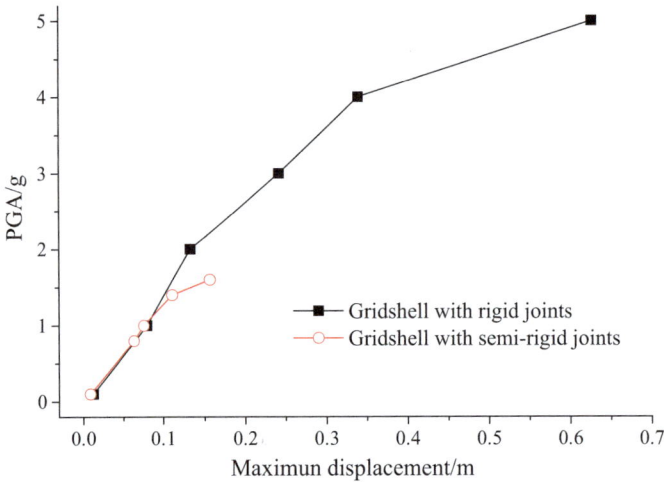

**Fig. 6.18** PGA-displacement curves for rigid-joint and semi-rigid-joint gridshells with 50 m span

## 6.6 Optimization Example 3: Single-Layer Gridshell with 80 m Span

### 6.6.1 Stable Optimization Considering Joint Stiffness

In Sect. 3.5, a single-layer K6 net shell with a span of 80 m and a chord height of 25 m is taken as an example and optimized for stability based on the assumption of rigid joints. The net shell is composed of six identical sectors, one of which is shown in bold in Fig. 6.19. Each sector consists of symmetrical red and black parts. The structure has a total of 2352 members and 817 joints. After the stability optimization in Sect. 3.5, the optimized structure has good stability performance, with $P_{cr} = 40.01$ kN/m². According to the JGJ7-2010 Code for Design of Space Grid Structures, the optimized structure in Sect. 3.5 is designed for joints. The joint form is a welded hollow ball joint, and the diameter and wall thickness of all hollow balls are taken as the minimum value that meets the strength requirements. After design, the steel used for the joints is 83,877 kg.

Using the stable optimization design method considering joint stiffness proposed in this chapter, the same single-layer shell structure in Sect. 3.5 is optimized for stability while determining the section and joints of the members. The relevant optimization parameters are as follows: the upper limit of steel used at the joints, $m_{j,0} = 26{,}000$ kg, in Eq. (6.26), the evolution rate of joint stiffness enhancement, $ei = 5\%$, in Eq. 6.27, and the evolution rate of joint stiffness reduction, $ed = 1\%$, in Eq. (6.28). The steel used in the members and other optimization parameters are consistent with those in Sect. 3.5.

## 6.6 Optimization Example 3: Single-Layer Gridshell with 80 m Span

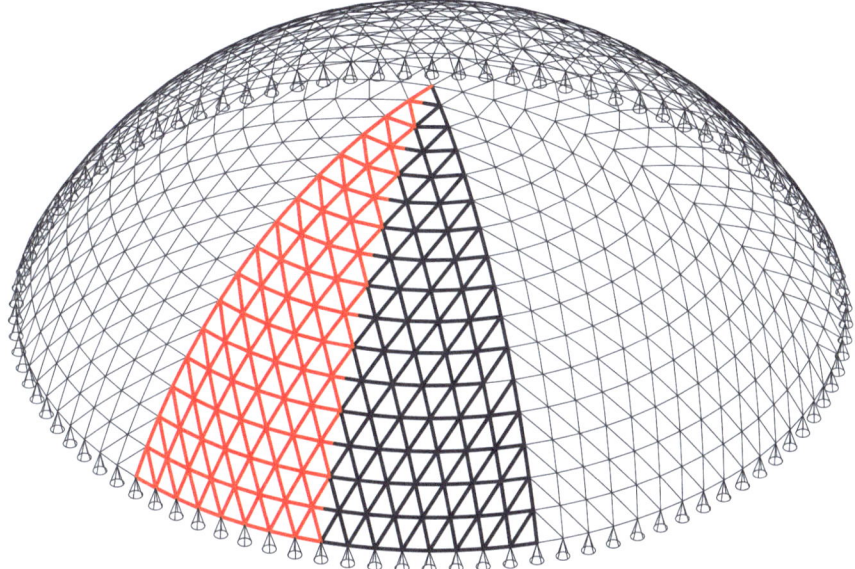

**Fig. 6.19** 80 m span single-layer spherical gridshell

The stable optimization algorithm considering nodal stiffness was run on a computer running the Windows 7 operating system, which was configured with an Intel(R) Core(TM) i7-4790 K CPU@ 4.00 GHz and 32 GB of memory. At the 120th optimization step, the population evolved to the optimal solution. The time required to evolve to the 120th optimization step was 4859.973 s, so the average time per generation was 40.450 s. The main steps of the stable optimization algorithm considering nodal stiffness and their corresponding times are shown in Table 6.12. For the single-story shell structure with an 80 m span, the time spent computing the structure's response accounted for 87% of the total time, which was approximately equal to the 50 m span shell structure in Sect. 6.5.2 (84%).

The optimization process of $P_{cr}$ and $gra\_r_{min}$ is shown in Fig. 6.20. $P_{cr}$ of the final optimized structure was 36.97 kN/m², $gra\_r_{min} = -8.449 \times 10^{-6}$. The steel quantity of the joint is 25910 kg $< m_{j,0}$, which meets the constraint requirement of the steel quantity of the joint. The amount of steel used for the rod is 45.46 < 46 kg/m², which meets the restriction requirements of the amount of steel used for the rod. Figure 6.20 shows that for this large-span single-layer latticed shell, the optimal design method can maximize gra_rmin, significantly reduce the degree of structural softening, and improve the structural stability bearing capacity.

After a stable optimization of the joint stiffness, the section distribution of the members is shown in Fig. 6.21. The elastic joint positions and numbers are shown in Fig. 6.22, and the specific characteristics of the elastic joints are shown in Table 6.13. Because the structure is symmetrical and the load is symmetrical, the member distribution is also symmetrical, so we will explain it using the 1/12 structure shown in red

**Table 6.12** Optimization algorithm considering joint stiffness for an 80 m span gridshell: main steps and time

| Main steps | Corresponding functions | Function overview | Time/s | Percentage of total time consumed (%) |
|---|---|---|---|---|
| Population initialization | InitialGen | Generating initial members | 0.718 | 0.01 |
| | Initial_Joint_stiffness | Initial joint stiffness | 0.047 | 0.00 |
| Decoding and encoding | GenToPhenotype | Generating binary code for membrane | 140.975 | 2.90 |
| | GAencode | Generating binary code | 0.078 | 0.00 |
| | GAdecode | Translating binary code | 11.811 | 0.24 |
| Joint shape degree calculation and individual fitness calculation | KKEE | Integrating overall stiffness matrix | 351.267 | 7.23 |
| | Boundary | Handling boundary conditions | 135.928 | 2.80 |
| | StressMatrix | Integrating geometric stiffness matrix | 393.186 | 8.09 |
| | MechanicsConstrain | Calculating mechanical constraint terms and penalty function | 282.085 | 5.80 |
| | FitnessCalculation | Calculating deformation gradient and individual fitness | 3064.514 | 63.05 |
| Oriented mutation of rods | find_variable | Identifying variables corresponding to the variant members | 2.763 | 0.06 |
| Oriented mutation of joints | Joint_Diameter_renew | Updating the diameter of the joint after variant members are altered | 6.606 | 0.14 |
| | Joint_stiffness_renew | Orienting the variant joint stiffness | 14.538 | 0.30 |
| File reading and writing | WriteBest | Outputting optimization results | 58.256 | 1.20 |
| Other | – | – | 397.201 | 8.17 |
| Total running time | – | – | 4859.973 | 100.00 |

## 6.6 Optimization Example 3: Single-Layer Gridshell with 80 m Span

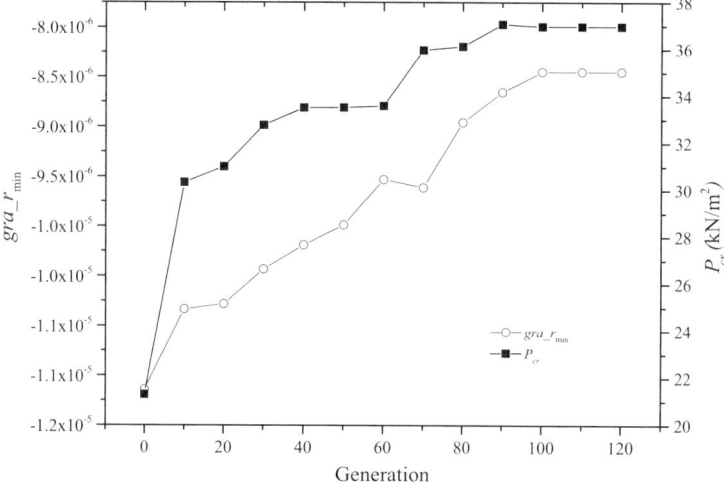

**Fig. 6.20** The optimization process of 80 m span mesh shell structure with $gra\_r_{min}$ and $P_{cr}$

in Fig. 6.19. Because the constraint condition is a fixed support around the perimeter, the joint at the support still uses the traditional welded hollow ball joint.

Considering the joint stiffness, the optimized netshell in this section has similar characteristics to the optimized netshell in Sect. 3.5, as shown in Table 6.14. Comparing the two optimized netshells reveals that, under the premise of nearly equal steel usage for the members, the stable bearing capacity of the semi-rigid joint optimized netshell structure is 92.4% of the stable bearing capacity of the rigid joint optimized netshell structure. The stable bearing capacities of the two optimized structures are almost equal, but the joint steel usage of the semi-rigid joint optimized netshell structure is only 35.18% of the joint steel usage of the rigid joint netshell structure.

### 6.6.2 The Anti-collapse Performance of Structures with Stable Optimized Stiffness After Joint Consideration

By comparing the collapse resistance performance of rigid joint truss structures and semi-rigid joint truss structures, the collapse resistance ability of the truss structure optimized for stability considering joint stiffness in Sect. 6.6.1 is verified. First, a numerical model based on the rigid joint assumption is established for the truss structure optimized in Sect. 3.5. In the numerical model, it is assumed that the joints are ideal rigidly connected, and other modeling techniques are the same as those in Sect. 6.4.2. After the rigid joint truss model is established, the numerical model of

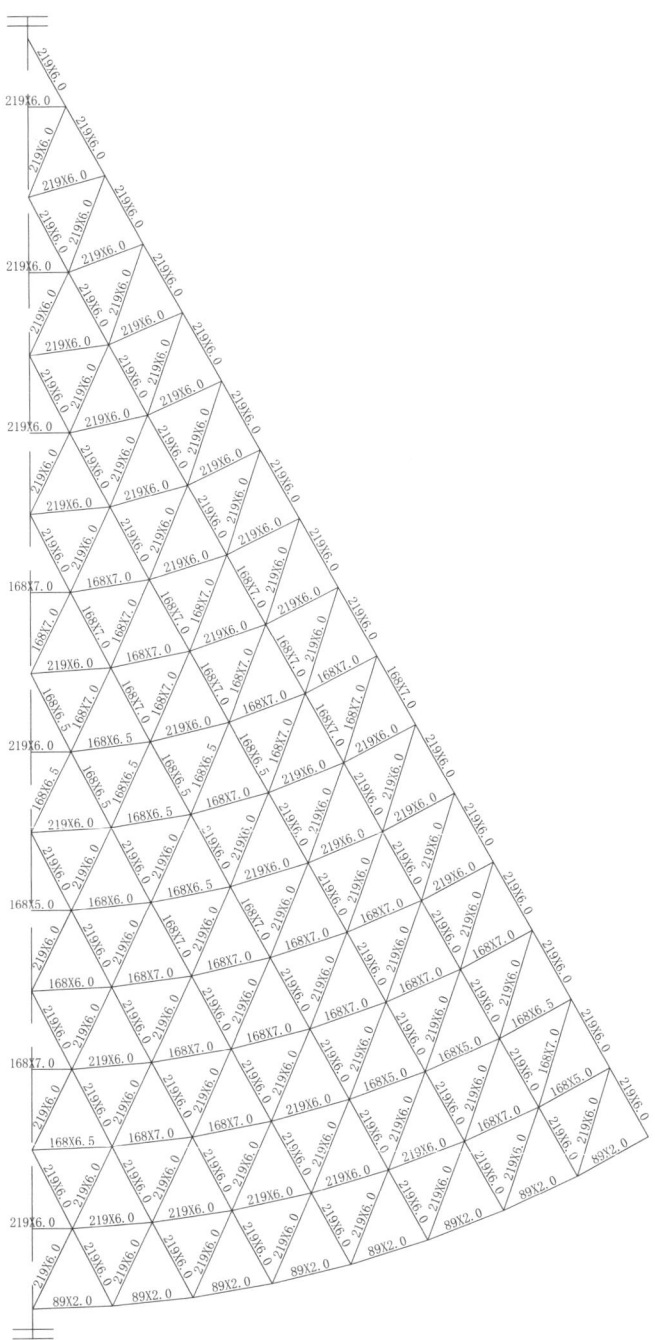

**Fig. 6.21** Sections of the optimized 80 m span single-layer gridshell considering joint stiffness

## 6.6 Optimization Example 3: Single-Layer Gridshell with 80 m Span

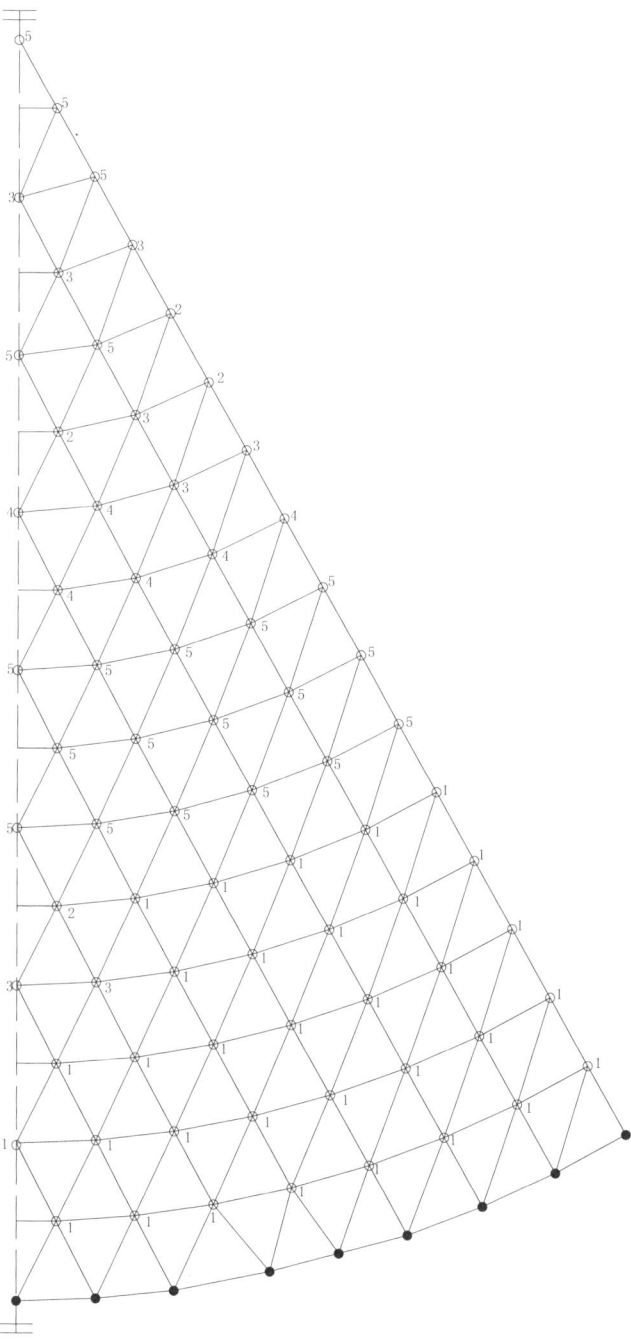

**Fig. 6.22** Joints of the optimized 80 m span single-layer gridshell considering joint stiffness

**Table 6.13** Joints of the optimized 80 m span single-layer gridshell

| Joint number | $P$/mm | m/kg | $k/10^6$ N m |
|---|---|---|---|
| 1 | 219 | 26.92 | 8.22 |
| 2 | 219 | 29.62 | 12.54 |
| 3 | 219 | 32.58 | 17.28 |
| 4 | 219 | 40.26 | 29.58 |
| 5 | 219 | 40.66 | 30.24 |

**Table 6.14** Optimized gridshells with 80 m span obtained by two optimization methods

|  | Optimized gridshell based on rigid joints | Optimized gridshell considering joint stiffness |
|---|---|---|
| $P_{cr}$ (kN/m$^2$) | 40.01 | 36.97 |
| gra_$r_{min}$ ($10^{-6}$) | − 7.924 | − 8.449 |
| Steel consumption for members (m$^3$) | 45.34 | 45.46 |
| Steel consumption for joints (kg) | 83,877 | 29,510 |

the truss structure optimized for stability considering joint stiffness in Sect. 6.5.1 is established. The yield moment of the optimized joints in Table 6.13 is determined according to the method in Sect. 6.4.2, so the joint stiffness and yield moment of the optimized truss structure in the collapse analysis are shown in Table 6.15. When establishing semi-rigid joint connections, a coordinate system is established at each semi-rigid connection point, and the stiffness and yield moment of the corresponding spring connection unit are accurately assigned.

Modal analysis was conducted on two cable net structures, and the first three natural frequencies of the structures are shown in Table 6.16. Table 6.16 indicates that: (1) The vibration frequencies of the semi-rigid joint cable net structure are lower than those of the rigid joint cable net structure, which is consistent with the concept of the structure; (2) The vibration frequencies of the semi-rigid joint cable net structure are slightly lower than the corresponding frequencies of the rigid joint cable net

**Table 6.15** Mechanical properties of semi-rigid joints in anti-collapse analysis of 80 m span shell structure

| Joint number | $P$/mm | m/kg | $k/10^6$ N m | $M^{yield}$/kN m |
|---|---|---|---|---|
| 1 | 219 | 26.92 | 8.22 | 15.398 |
| 2 | 219 | 29.62 | 12.54 | 17.634 |
| 3 | 219 | 32.58 | 17.28 | 20.088 |
| 4 | 219 | 40.26 | 29.58 | 26.450 |
| 5 | 219 | 40.66 | 30.24 | 26.780 |

## 6.6 Optimization Example 3: Single-Layer Gridshell with 80 m Span

structure at the same order, indicating that the optimization algorithm selected the joints without significantly reducing the dynamic performance of the structure in the linear elastic stage; (3) The frequency distribution patterns of the two structures are the same, verifying the numerical model of the semi-rigid joint cable net structure.

The same three-dimensional El-centro wave input as in Sect. 6.4.2 was used to perform nonlinear time history analyses of the two structures, step by step increasing the seismic ground motion PGA. The load–displacement curves for the two structures are shown in Fig. 6.23. The maximum PGA that the rigid joint truss structure can withstand is 3.4 g. The load–displacement curve becomes increasingly flat as the PGA is increased, indicating that the optimized truss structure based on the rigid joint assumption experiences nonlinear deformation under seismic action, with a predictable collapse failure mode as the PGA increases. The maximum PGA that the semi-rigid joint truss structure can withstand is 1.0 g. When PGA is less than 0.7 g, the maximum deformation of the semi-rigid joint truss structure is almost the same as that of the rigid joint truss structure, which is verified by the modal analysis results. When PGA is greater than 0.7 g, the curvature of the load–displacement curve of the semi-rigid joint truss structure becomes flatter than that of the rigid joint truss structure before it becomes increasingly flat. The collapse critical load of the semi-rigid joint truss structure is 1.0 g, which is smaller than the collapse critical load of the rigid joint truss structure at 3.4 g. The reasons for the reduction of the collapse critical load of semi-rigid joint shell structures are as follows: (1) The M-θ relationship (see Fig. 6.12) of the joints is a conservative model. Firstly, when the maximum stress at the joint reaches the yield stress, the conservative assumption is made that the moment at the joint at this time is the yield moment of the joint. Secondly, when the joint enters the plastic state, the plastic development is no longer considered, and the slope of the M-θ curve is conservatively set to zero. Meanwhile, the model has not yet considered the dynamic characteristics such as hysteretic behavior of the joints; (2) In this section, the ratio of the steel used in the joints to the steel used in the members is 0.129:1, which is clearly smaller than the ratio in Sect. 6.4, and is basically the same as that in Sect. 6.5, so that some joints yield before the members, forming plastic hinges; (3) In this section, the small span-to-chord ratio of the shallow arch and the spherical curvature radius of 44.5 m make the shell surface the flattest, so that the shell is more sensitive to vertical seismic motion. After the plastic hinge is formed, it is easier to form the structure in three-way reciprocating seismic action, causing the original structure's configuration to be destroyed. Despite the fact that the collapse critical load of semi-rigid cable net structures is lower than that of rigid-jointed cable net structures, the performance of semi-rigid-jointed cable net structures with a 9° rare earthquake (0.62 g) is basically the same as that of

**Table 6.16** First three frequencies of optimized gridshells with two 80 m spans

|  | 1st frequency/Hz | 2nd frequency/Hz | 3rd frequency/Hz |
|---|---|---|---|
| Ideal rigidly connected gridshell | 4.4521 | 4.4521 | 5.3519 |
| Semi-rigidly connected gridshell | 4.3672 | 4.3672 | 5.2095 |

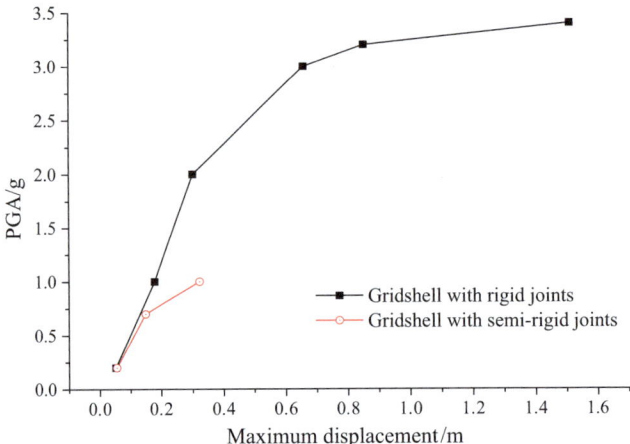

**Fig. 6.23** PGA-displacement curves for rigid-joint and semi-rigid-joint tensile gridshells with 80 m span

rigid-jointed cable net structures, and its collapse critical load has far exceeded the peak acceleration value of the 9° rare earthquake (0.62 g), and its collapse precursor is more obvious.

## 6.7 Chapter Summary

The classical configuration fragility theory assumes that the joints are ideal rigidly connected, and cannot consider the influence of joint stiffness. To overcome this deficiency, this chapter further extends the configuration fragility theory by modifying the element stiffness matrix to consider the influence of joint stiffness, thereby revealing the stability of semi-rigid joint single-layer shell structures. Then, taking a single-layer shell structure as an example, with $P_{cr}$ and $gra\_r_{min}$ representing the structural stability, the quantitative study of the influence of joint stiffness on the stability performance of single-layer shell structures was conducted, verifying that the configuration fragility theory considering joint stiffness can reveal the stability of semi-rigid joint shell structures, and clarifying the reasonable range of joint rotational stiffness. Based on the reasonable range of joint rotational stiffness, combined with the joint optimization design method presented in Chap. 5, the optimized joint library was obtained for the joints within the rotational stiffness range.

For semi-rigid joint single-layer shell structures, a stable optimization design method considering joint stiffness was proposed. In the optimization model, the degree of structural stiffness degradation was represented by the computationally simple gra_rmin, which was used to measure the structural stable bearing capacity. The optimization goal was to reduce the softening degree of the structure, thereby improving the structural stable bearing capacity. Both the members and the joints were the optimization variables. The member section was taken from the Chinese manufacturing standard GB/T 17395-2008 Dimensions, Shapes, Weights and Permissible Deviations of Seamless Steel Tubes, and the joint variable was taken from the optimized joint library. The design requirements stipulated by the Chinese design code were used as the optimization constraint conditions. The optimization model can conduct stable optimization design of single-layer shell structures under the premise of giving upper limits on the steel usage of members and joints. An optimization algorithm was developed for the case where members and joints are both optimization variables in the optimization model. Three single-layer shell structures with different spans were used as examples to verify the stable optimization design method considering joint stiffness. The stable optimization results of the three different span structures all show that the stable optimization design method considering joint stiffness can determine both members and joints at the same time, and improve the structural stable bearing capacity under the premise of very economical steel usage of members and joints; Compared with the single-layer shell structure stabilized by rigid joint optimization, the single-layer shell structure stabilized by considering the joint stiffness optimization can significantly reduce the steel usage at joints while maintaining the stability carrying capacity unchanged. When the joint transitions from ideal rigid connection to semi-rigid connection, to verify the seismic anti-collapse performance of the semi-rigid joint shell structure, the above three semi-rigid joint optimized shell structures are taken as examples and compared with the corresponding rigid joint shell structure's anti-collapse performance. The calculation results show that the frequency of the semi-rigid joint shell structure is basically the same as that of the rigid joint shell structure, and its maximum displacement response in a 9° rare earthquake is basically the same as that of the rigid joint shell structure; the collapse critical load of the semi-rigid joint shell structure is lower than that of the rigid joint shell structure, but it is still significantly higher than the peak acceleration value of the 9° rare earthquake motion specified in the seismic code.

# References

1. Monforton GR (1962) Matrix analysis of frames with semi-rigid connections. University of Windsor
2. Han Q, Liu Y, Xu Y (2016) Stiffness characteristics of joints and influence on the stability of single-layer latticed domes. Thin-Walled Struct 107:514–525
3. National Standard of the People's Republic of China (2008) GB/T 17395–2008 dimension, shape, weight and allowable deviation of seamless steel pipes. Standards Press of China, Beijing
4. National Standard of the People's Republic of China (2003) GB 50017–2003 code for design of steel structures. China Architecture and Building Press, Beijing
5. Industrial standard of the People's Republic of China (2010) JGJ7-2010 technical specification for space grid structure. China Architecture and Building Press, Beijing
6. Adeli H, Cheng NT (1994) Concurrent genetic algorithms for optimization of large structures. J Aerosp Eng 7(3):276–296

**Open Access** This chapter is licensed under the terms of the Creative Commons Attribution-NonCommercial-NoDerivatives 4.0 International License (http://creativecommons.org/licenses/by-nc-nd/4.0/), which permits any noncommercial use, sharing, distribution and reproduction in any medium or format, as long as you give appropriate credit to the original author(s) and the source, provide a link to the Creative Commons license and indicate if you modified the licensed material. You do not have permission under this license to share adapted material derived from this chapter or parts of it.

The images or other third party material in this chapter are included in the chapter's Creative Commons license, unless indicated otherwise in a credit line to the material. If material is not included in the chapter's Creative Commons license and your intended use is not permitted by statutory regulation or exceeds the permitted use, you will need to obtain permission directly from the copyright holder.

# Chapter 7
# Conclusion and Prospects

**Abstract** At present, the design method of single-layer gridshells is based on the assumption of rigid joints, and the stability problem, which dominates the structural design, is considered by verification. The design-verification-redesign iterative method will reduce the design efficiency, and the rigid joints hinder the application of new semi-rigid joints, which is not conducive to the promotion of prefabricated buildings.

## 7.1 Main Conclusions

At present, the design method of single-layer gridshells is based on the assumption of rigid joints, and the stability problem, which dominates the structural design, is considered by verification. The design-verification-redesign iterative method will reduce the design efficiency, and the rigid joints hinder the application of new semi-rigid joints, which is not conducive to the promotion of prefabricated buildings. To solve these problems, this paper conducts a systematic study in the following aspects: revealing the instability mechanism of gridshells, stability optimization design method, verification of stability optimization design method, new joint development, and stability optimization design method considering joint stiffness. The main innovations and conclusions are as follows:

1. By introducing external factors such as loads and supports into the classical damageability theory in the form of geometric stiffness matrix, the limitation of the classical damageability theory in not considering loads is solved, expanding the applicability of the damageability theory. Furthermore, the newly expanded damageability theory is used to reveal the instability mechanism of single-layer shell structures from a new perspective. The following conclusions are obtained:

    (1) Under a given loading pattern, analyzing the changes in the nodal geometric degree of freedom (i.e., $gra\_r$) before and after loading can quantitatively measure the degree of stiffness degradation of the joints. The joint with the minimum $gra\_r$ (i.e., $gra\_r_{min}$) has the most significant softening, loses stability first during the loading process, and is the key joint that determines

the structural stable bearing capacity. The joint district with lower *gra_* values has obvious softening, and local instability occurs during the loading process. When the *gra_* values of all joints are similar, it indicates that the structure as a whole is softening, and the instability occurs in the global instability mode.

(2) Under different loading modes, comparing the distribution of *gra_r* and *gra_r*$_{min}$ at the joints can determine the most unfavorable loading mode for the stable angle. For the K6 spherical shell structure, under the concentrated load mode, only the shape degradation of the load-bearing points is significant, indicating that the structure's resistance to load has not yet been fully utilized, and this loading mode is the most unfavorable loading mode. Under the semi-span uniformly distributed load mode, the shape degradation of the load-bearing joints in the load zone is significantly higher than that in the non-load zone, and the stable bearing capacity is higher than that under the concentrated load at the peak. Under the full-span uniformly distributed load, the *gra_r* distribution of all joints is uniform and the peak value is low, indicating that the structure is loaded reasonably, and the corresponding stable bearing capacity is the highest.

2. Based on the instability mechanism and rigid joint assumption of the gridshell, a stable optimization model for single-layer gridshell is established. For the problem of large number of optimization variables and difficulty in fast solution of traditional optimization algorithms in the stable optimization model, the random mutation mechanism in the standard genetic algorithm is improved into a directed mutation mechanism, and a guided genetic algorithm is proposed to improve search efficiency. Three single-layer gridshells with different spans are used as examples to verify the stable optimization method. The following conclusions are obtained:

(1) The optimization model optimizes the structure's stability from the perspective of structural instability, using a simple *gra_r*$_{min}$ representation to calculate the degree of structural stiffness degradation; the optimization goal is to reduce the structure's softening degree to improve its stable bearing capacity; the optimization variables are discrete beam section dimensions; and the design limit values specified by codes are the constraints. The optimization model can significantly reduce the structure's instability trend under the premise of a maximum steel usage of the individual beam elements, achieving optimized anti-instability design for single-layer shell structures. For the 22 m small-span stable optimization shell structure, $P_{cr} = 94.22$ kN/m$^2$ under the condition that the steel usage per unit area does not exceed 29.31 kg/m$^2$, which is 27.9% higher than the initial structure; for the 50 m medium-span stable optimization shell structure, $P_{cr} = 32.55$ kN/m$^2$ under the condition that the steel usage per unit area does not exceed 24.46 kg/m$^2$, which is 22.1% higher than the initial structure; for the 80 m large-span stable optimization shell structure, $P_{cr} = 40.01$ kN/m$^2$ under the condition

that the steel usage per unit area does not exceed 46 kg/m², which is 30.5% higher than the initial structure.

(2) For small span cable net structures, the standard genetic algorithm takes 780 min to obtain the optimized solution, while the guided genetic algorithm only takes 11 min. For medium span and large span cable net structures, the standard genetic algorithm is unable to solve the problem, while the guided genetic algorithm takes 73 and 110 min respectively to evolve to the optimized solution, and both converge to the optimized solution. The guided genetic algorithm can quickly solve stable optimization models, with short computation time and reliable and stable optimization results.

3. Two large single-layer shell structures were subjected to stability optimization to further verify the stability optimization design method. Using numerical simulation methods based on the assumption of rigid joints, the strong earthquake anti-collapse analysis of the optimized structures was conducted, and the results were compared with the initial structure's shaking table collapse test. The following conclusions were obtained:

(1) The number of members in the large single-layer shell structure reaches 3660, and the number of candidate cross-sections for each optimization variable ranges from 575. For the two large-scale stability optimization problems, the stability optimization design method can still increase the $P_{cr}$ of the two initial structures by 1.732 times and 1.812 times respectively, verifying the feasibility of the stability optimization design method.

(2) The support conditions, geometric dimensions, topological connections, and steel usage of the two initial shells are identical, with only slightly different initial member cross-sections. However, the corresponding optimized structures have similar $gra\_r_{min}$ values, and $P_{cr}$ values are basically the same, with the same instability patterns. This verifies the robustness of the optimization algorithm.

(3) By numerical simulation, the load–displacement curves of the optimized structure were obtained, and they were compared with the load–displacement curves obtained from the shaking table test of the initial structure. The results show that the structure optimized for stability not only has good static stability performance, but also the peak acceleration value of the critical earthquake motion before collapse of the optimized structure is 2 times higher than that of the initial structure, and the pre-collapse precursor is significant, showing good seismic resistance to collapse.

4. To enrich the joint forms of single-layer shell structures, this paper optimizes the joints of single-layer shell structures from the perspectives of improving the joint rotational stiffness and enhancing the joint safety. To improve the joint rotational stiffness, a topology optimization model for the joints is established, and a solution is proposed to the technical difficulty of how to apply equivalent joint concentrated forces. At the same time, the design of spherical joint ends is combined with construction design. To enhance the safety performance of

the joints, two two-dimensional joints are optimized based on structural safety evaluation indicators. The following conclusions are drawn:

(1) In the topology optimization model of joint stiffness, the joint stiffness is represented by the self-balancing force system applied, which avoids the dependency of the optimization result on the loading condition. Based on this joint stiffness, the topology optimization model of maximum joint stiffness with fixed mass and the topology optimization model of minimum mass with fixed stiffness are established respectively.

(2) In the joint stiffness topology optimization model, the inertial force is applied to achieve the balance of joints by equivalent joint concentrated load, which avoids the disadvantages of stress singularity and joint form limitation caused by direct application of concentrated force on the design domain, and also avoids the support providing the balancing force, so that the boundary condition is only to limit the rigid body displacement, thus obtaining stable and reliable optimization results.

(3) Using topology optimization algorithm to solve the joint stiffness optimization model, a reasonable structure form, lightweight, and high-stiffness joint core was obtained. The optimized joint core consists of radially connected elements and inner ring hubs. Under the action of out-of-plane bending moment, the radially connected elements transmit the force to the ring hub, and the ring hub resists the load almost throughout its cross-section. Under the action of in-plane bending moment, the radially connected elements and the ring hub form a plane arch to transmit and balance the end force, while the plane arch supports each other while maintaining the flow of force without interruption, thereby improving the structural stiffness. Compared with the welded hollow ball joint, the minimum optimized joint stiffness is 88.7% of the welded hollow ball joint, but the out-of-plane stiffness of the optimized joint is 1.8 times that of the welded hollow ball joint, and the in-plane stiffness is 2.3 times that of the welded hollow ball joint.

(4) For the optimized topology joints, a spherical joint end is designed by combining the construction requirements of prefabricated buildings. The spherical joint end makes the joint directionless, has good geometric adaptability, good member adaptability, and has an auxiliary alignment function.

(5) To enhance the safety performance of joints, an independent structural safety evaluation index is proposed, which is not dependent on the load. Based on this index, a topological optimization model for joint safety performance is established. The topological optimization of two two-dimensional joints is carried out to verify the joint safety optimization design method.

5. To consider the influence of joint stiffness in the stability optimization model, this paper further extends the theory of geometric fragility and proposes a stability optimization design method considering joint stiffness. The stability optimization design method considering joint stiffness takes both the members and joints as optimization variables. The stability optimization design method considering

## 7.1 Main Conclusions

joint stiffness is verified by using three single-layer shell structures with different spans as examples. The following conclusions are obtained:

(1) By correcting the element stiffness matrix to take into account the influence of joint stiffness, the limitation of the classical configuration fragility theory that cannot consider the influence of joint stiffness is overcome, further expanding the scope of the configuration fragility theory, enabling it to reveal the instability mechanism of semi-rigid joint single-layer shell structures.

(2) For example, a single-layer gridshell is studied quantitatively by using $P_{cr}$ and $gra\_r_{min}$ to represent the structural stability, and the influence of joint stiffness on the stability performance of single-layer gridshell is analyzed. The reasonable range of joint rotational stiffness is clarified. Within the reasonable rotational stiffness range, the optimal design of different-sized joints is carried out, and an optimized joint library is obtained.

(3) In the stable optimization design method, the joint stiffness is taken into account, with both joints and members being optimized variables. The cross-section of members is taken from the manufacturing standard of our country, and the joints are taken from the optimized joint library. An optimization model for single-layer shell structure considering joint stiffness is established with the constraints of steel usage for members, steel usage for joints, and design limits, and an efficient optimization algorithm is developed.

(4) Select three single-layer ball shell initial structure examples with large, medium, and small spans, whose control parameters include span, aspect ratio, geometric shape, total steel usage, etc., which are the same as the stable optimization examples of rigid joint truss shells. Conduct stable optimization design considering joint stiffness. The results show that the 22 m small span semi-rigid joint stable optimization truss shell has $P_{cr} = 92.55$ kN/m$^2$, which is 2.03 times higher than the initial structure, and is basically equal to the $P_{cr}$ of the corresponding rigid joint stable optimization truss shell, but its joint steel usage is 40% lower than that of the corresponding rigid joint optimized truss shell. The 50 m medium span semi-rigid joint stable optimization truss shell has $P_{cr} = 32.11$ kN/m$^2$, which is 1.52 times higher than the initial structure, and is basically equal to the $P_{cr}$ of the corresponding rigid joint stable optimization truss shell, but its joint steel usage is 66.6% lower than that of the corresponding rigid joint optimized truss shell. The 80 m large span semi-rigid joint stable optimization truss shell has $P_{cr} = 36.97$ kN/m$^2$, which is 1.74 times higher than the initial structure, and is basically equal to the $P_{cr}$ of the corresponding rigid joint stable optimization truss shell, but its joint steel usage is 64.8% lower than that of the corresponding rigid joint optimized truss shell.

(5) The seismic resistance against collapse of semi-rigid joint stable optimization structural systems and rigid joint stable optimization structural systems under strong earthquakes is analyzed based on the assumption of rigid joint. From the load–displacement curves of each system, it can be found that the critical collapse load of the semi-rigid joint gridshell is lower than that

of the rigid joint gridshell, but still higher than the peak acceleration value of the 9th degree rare earthquake motion stipulated in the seismic code of our country, indicating that the semi-rigid joint stable optimization gridshell still has good seismic resistance against collapse performance.

## 7.2 Future Outlook

This paper addresses the problems existing in the current single-layer dome structure design method and proposes a stable optimization design method for single-layer dome structures, which is conducive to improving design efficiency and promoting and using new types of joints. At the same time, there are some areas that need to be further explored:

(1) Firstly, this paper reveals the instability mechanism of the gridshell from the perspective of structural mathematical model (i.e. structural stiffness matrix), breaking the limitation of structure form, and the conclusion obtained is applicable to all single-layer gridshells. Secondly, in the stable optimization algorithm, the method proposed in this paper is not dependent on the designer's experience and is not limited to specific structure forms, and the optimization design method can be applied to all single-layer gridshells. In the revelation of the instability mechanism of the gridshell, the relevant examples in this paper are all single-layer K6 gridshells. The next step is to extend the stable optimization design method proposed in this paper to other types of single-layer gridshells. Meanwhile, the instability mechanism revealed in this paper is also applicable to double-layer gridshells, and the stable optimization design method can be extended to double-layer gridshells as well.

(2) In the research and development of new joint types, this paper only proposes the mechanical model of the K6 type truss shell. For other types of single-layer truss shell structures, corresponding joint mechanical models need to be established. At the same time, this paper only proposes a connection interface for circular tube members. For other types of member cross-section shapes, further research is needed on the connection interface. At the same time, it is necessary to further refine the design of the connection in combination with the requirements of prefabricated construction, and propose specific and feasible connection technologies.

(3) The new joint designs were obtained through topology optimization, with unique shapes that can be combined with advanced manufacturing methods such as 3D printing to achieve the cost-effective and reliable industrial production of new joint designs.

(4) In the evaluation and optimization of joint safety, the selection of reasonable evaluation indicators should not only be able to reflect the essence of safety deeply, but also be convenient for calculation in the optimization process. Although the safety evaluation indicators selected in this paper can truly reflect

## 7.2 Future Outlook

the safety performance of joints, the calculation process involves matrix inversion, so the calculation volume increases exponentially with the increase of the structure scale. Therefore, other indicators can be selected for reference, following the idea of safety evaluation indicators in this paper, which can reasonably evaluate the joint safety and are convenient for calculation, thus achieving the optimization of the safety performance of three-dimensional joints.

**Open Access** This chapter is licensed under the terms of the Creative Commons Attribution-NonCommercial-NoDerivatives 4.0 International License (http://creativecommons.org/licenses/by-nc-nd/4.0/), which permits any noncommercial use, sharing, distribution and reproduction in any medium or format, as long as you give appropriate credit to the original author(s) and the source, provide a link to the Creative Commons license and indicate if you modified the licensed material. You do not have permission under this license to share adapted material derived from this chapter or parts of it.

The images or other third party material in this chapter are included in the chapter's Creative Commons license, unless indicated otherwise in a credit line to the material. If material is not included in the chapter's Creative Commons license and your intended use is not permitted by statutory regulation or exceeds the permitted use, you will need to obtain permission directly from the copyright holder.

MIX
Papier aus verantwortungsvollen Quellen
Paper from responsible sources
FSC® C105338

If you have any concerns about our products,
you can contact us on
**ProductSafety@springernature.com**

In case Publisher is established outside the EU,
the EU authorized representative is:
**Springer Nature Customer Service Center GmbH
Europaplatz 3, 69115 Heidelberg, Germany**

Printed by Libri Plureos GmbH
in Hamburg, Germany